Comparative Pathobiology

Volume 8

PARASITIC AND RELATED DISEASES

Basic Mechanisms,
Manifestations, and Control

Comparative Pathobiology

Comparative Pathobiology

Volume 8

PARASITIC AND RELATED DISEASES
Basic Mechanisms, Manifestations, and Control

Edited by Thomas C. Cheng

Medical University of South Carolina
Charleston, South Carolina

Plenum Press • New York and London

Library of Congress Cataloging in Publication Data

Main entry under title:

Parasitic and related diseases.

(Comparative pathobiology; v. 8)
"Selected papers from the symposia of the American Society of Zoologists, held
December 27–30, 1983 in Louisville, Kentucky; the Society for Invertebrate Pathology,
held August 6–11, 1983, at Cornell University, Ithaca, New York; and the American
Society of Parasitologists, held September 4–8, 1983, in San Antonio, Texas" — T.p.
verso.
Includes bibliographies and index.
1. Parasitology — Congresses. 2. Parasitic diseases — Congresses. 3. Invertebrates —
Immunology — Congresses. 4. Physiology, Pathological — Congresses. I. Cheng,
Thomas Clement. II. American Society of Zoologists. III. Society for Invertebrate
Pathology. IV. American Society of Parasitologists. V. Series.
QL757.P28   1985   591.2′3   85-24447
ISBN 978-1-4684-5029-3   ISBN 978-1-4684-5027-9 (eBook)
DOI 10.1007/978-1-4684-5027-9

Selected papers from the symposia of the American Society of Zoologists,
held December 27–30, 1983, in Louisville, Kentucky; the Society of
Invertebrate Pathology, held August 6–11, 1983, at Cornell University,
Ithaca, New York; and the American Society of Parasitologists,
held September 4–8, 1983, in San Antonio, Texas

© 1985 Plenum Press, New York
Softcover reprint of the hardcover 1st edition 1985
A Division of Plenum Publishing Corporation
233 Spring Street, New York, N.Y. 10013

The study of parasites and their interactions with hosts
continues to represent a challenging area of modern biology.
The availability of new techniques and instrumentation, coupled
with the development of daring new hypotheses and concepts, has
paved the way for the dramatic evolution of parasitology from a
static descriptive endeavor to a dynamic one based on biochemistry,
immunology, molecular biology, and modern cell biology. Studies
of this nature obviously fall within the domain of pathobiology.
Consequently, when the contributions included in this volume
of *Comparative Pathobiology* were offered to this series, after
critical review, we welcomed the opportunity to make them available
to the scientific community.

The contributions included herein represent presentations
delivered before enthusiastic audiences at three different
symposia, all held in 1983. The first, entitled "Some Aspects
of Modern Parasitology", was organized by Dr. Gary E. Rodrick of
the University of South Florida and myself on behalf of the
American Society of Zoologists. The chapters by C. E. Carter and
B. M. Wickwire, B. J. Bogitsh, and W. M. Kemp were originally
presented at that symposium.

The second symposium, organized by Dr. G. Balouet of the
Faculté de Médecine, Brest, France, and myself on behalf of the
Society for Invertebrate Pathology, was entitled "Cellular Reactions
in Invertebrates." The chapters by G. Balouet and M. Poder and
M. Brehelin were originally presented at this symposium.

The third event, designated as the "President's Symposium" by
Dr. Donald Heyneman, past-president of the American Society of
Parasitologists, was convened in honor of two retiring medical
malacologists, Dr. Charles S. Richards of the National Institutes
of Health, Bethesda, Maryland, and Dr. Kian Joe Lie of the
University of California Medical Center, San Francisco. The
symposium was organized by Dr. Christopher J. Bayne of Oregon State
University. The contributions by D. S. Woodruff and T. C. Cheng
in this volume were originally presented at this symposium.

Finally, Dr. D. S. Woodruff offered the contribution by M.
Fletcher for inclusion in this volume because of the obvious

relatedness of the subject material with that of his chapter and
the general theme of this volume, and Dr. A. Mohandas offered
his contribution for inclusion because of the relatedness of his
work to that of G. Balouet and M. Poder, M. Brehelin, and T. C.
Cheng. After critical review, it was decided that the inclusion of
these chapters would enrich this volume.

It is gratifying to know that *Comparative Pathobiology* is
reaching a large audience, the contents of the first seven volumes
are widely quoted, and we have assisted in defining the role of
pathobiology in modern biology.

As has been stated in earlier volumes, those wishing to sub-
mit manuscripts clustered around a central theme relevant to
pathobiology should contact the editor.

                              Thomas C. Cheng
                              Charleston, South Carolina

CONTENTS

SELECTED ASPECTS OF ENZYME REGULATION IN PARASITES

Clint E. Carter and Brian M. Wickwire

Department of Biology
Vanderbilt University
Nashville, Tennessee 37235

I. SYNOPTIC REVIEW

Carbohydrate has been shown to be the major energy reserve in parasitic helminths. The details of carbohydrate degradation have been studied in a number of helminths (Bueding and Saz, 1968; Saz, 1970, 1971). It is clear from these studies that carbohydrate catabolism results in the production of acid end products which are persistent under either aerobic or anaerobic conditions. Carbohydrate catabolism in these helminths follows the conventional glycolytic sequence to phosphoenol-pyruvate (PEP) which may then be acted upon by pyruvate kinase (2.7.1.40, ATP: Pyruvate phosphotransferase) to yield pyruvate and subsequently, lactate and acetate, or by phosphoenolpyruvate carboxykinase (4.1.1.32, GTP: oxalacetate carboxylase) to yield oxaloacetate, malate, and ultimately succinate (Saz 1970, 1971).

Based on the major fermentation end product released, there appear to be three major categories of glucose fermentation in parasitic helminths. The first group is represented by the nematode *Ascaris lumbricoides* and is characterized by the production of succinate. The second group is represented by the adult cestode *Hymenolepis diminuta* which produces a combination of succinate, lactate, and acetate in a molar ratio of 5.1:2.5:1.

The third group is represented by the homolactate fermenter
Schistosoma mansoni.

Bueding and Saz (1968) have suggested that the relative pro-
portions of fermentation acids derived from pyruvate or oxaloacetate
can be determined by competition between pyruvate kinase and PEP
carboxykinase for the same substrate, PEP.

As an example, Table 1 shows the ratio of Pyk/PEPCK specific
activity from three different helminths. Pyruvate kinase
activity is difficult to detect in adult *Ascaris* muscle and it is
at least twenty-five times lower than the activity of PEP carboxykinase (Bueding and Saz, 1968). Under these conditions, it is clear
that the flow of PEP is shunted toward the production of succinate
and its derivatives.

TABLE 1. Ratio of Pyk/PEPCK specific activity in three helminths.

SOURCE	Pyk/PEPCK	REFERENCES
Ascaris Body Muscle	0.04	Beuding & Saz, 1968
H. diminuta	0.28	Carter & Fairbairn, 1975
S. mansoni	5.0 - 9.7	Beuding & Saz, 1968

By 15 days post-infection, *H. diminuta* has developed into a
sexually mature adult. The adult *H. diminuta* represents a con-
tinuous state of development along the length of the strobila.
This developmental sequence includes a scolex, the neck region
where the proglottids originate, and the immature, sexually mature,
and gravid proglottids. Watts and Fairbairn (1974) have demonstrated that the adult worms excrete succinate, acetate, and lactate in
a molar ratio of 5.1:2.5:1.0, while 6-day-old immature worms
secrete the same acids with a molar ratio of 0.9:0.6:1.0. When
the ratio of Pyk/PEPCK activity in the various developmental stages
of *H. diminuta* was examined, it was clear that there is quantitative
shift during development from acids depending upon pyruvate kinase
activity to acids depending on PEP carboxykinase activity (Carter
and Fairbairn, 1975).

Pyruvate kinase from mammalian systems is known to exist in at least three isozymic forms. Pyk I (or type L) is present in liver and erythrocytes, Pyk II (or type LM) found in the kidney, and Pyk III (or type M) present in muscle and several other tissues (Tanaka et al., 1967; Jiminex de Asua et al., 1971). All forms of the enzyme require both divalent, magnesium or manganese, and mono-valent cations, potassium or ammonium, along with ADP and PEP for activity (Jacobson and Black, 1971). Pyk I is allosterically modulated by PEP, K^+, fructose 1,6-diphosphate, ATP, and alanine (Carminatti et al, 1971; Vivayuarigiya et al., 1969). Pyk III gives a hyperbolic velocity curve with respect to PEP, and its activity is not affected by fructose-1,6-diphosphate. The properties of Pyk II appear to be intermediate between Pyk I and III. The different kinetic properties of these isozymes are thought to re-flect different functions and requirements for control of the glycolytic pathway in various tissues.

Substrate (PEP) concentration vs. velocity curves for total kinase activity from homogenates of adult *H. diminuta* result in sigmoidal profiles with multiple plateaus. However, the positive allosteric effector fructose-1,6-diphosphate converts this sig-moidal curve to the hyperbolic curve associated with Michaelis-Menten kinetics (Carter and Fairbairn, 1975). At least five dis-tinct isozymes of Pyk can be separated from this homogenate of adult worms by DEAE-cellulose chromatography. Each of these isozymes display their own specific kinetic parameters. To summarize the data published by Carter and Fairbairn (1975), four of the five isozymes appear similar to the liver (type I) in that their $K0.5_{(PEP)}$ are relatively high and they are allosterically modulated by FDP. K_m values for the activated isozymes are from 4-25x lower than their $K0.5_{(PEP)}$ values. Based on the data of Bueding and Saz (1968), the concentration of PEP and pyruvate in adult *H. diminuta* are both about 4.0mM. This concentration is lower than the $K0.5_{(PEP)}$ of the nonmodulated forms of the enzymes but is within the range after activation by FDP. One of the five isozymes identified from adult worm extracts of *H. diminuta* appears similar to the muscle form (type III) as its $K_{m(PEP)}$ is low and not activated by FDP. The data of Barrett and Beis (1972) indicate that the concentration of ADP in *H. diminuta* exceeds the $K_{m(ADP)}$ in all the isozymes. These data indicate that the competition between pyruvate kinase and PEP carboxykinase is very likely controlled by fructose-1,6-diphosphate concentrations. As has been discussed by Carter and Fairbairn (1975), *H. diminuta* is an obligatory fermenter in which gluconeogenesis is minimal. Hence, the probable function of these modulated isozymes is to regulate the specific composition of lactate, acetate, and succinate excreted at different stages of development.

The early work on carbohydrate metabolism of *S. mansoni* by Bueding (1950) clearly demonstrated that adult worms are homolactate fermenters capable of utilizing glucose in quantities equivalent to 1/5 of their dry weight/hr *in vitro*. As is to be expected, the Pyk/PEPCK ratio reported by Saz and Bueding (1968) is 5.0-9.7. An even higher ratio of 30.3 has been reported by Brazier and Jaffe (1972). This indicates strongly that in schistosomes, the flux from the PEP branchpoint is controlled mainly by pyruvate kinase. Brazier and Jaffe (1972) reported that *S. mansoni* Pyk closely resembles the muscle form of pyruvate kinase in that it is insensitive to FDP activation and is relatively insensitive to ATP inhibition. This could be considered consistent with homolactate fermentation displayed by the parasite.

However, in some recent experiments conducted in our laboratory, considerably different results were obtained with the pyruvate kinase from *S. japonicum*. As a result of these experiments, a comparative study was undertaken between *S. mansoni* and *Schistosoma japonicum*.

Pyruvate kinase from *S. japonicum* gives with increasing phosphoenolpyruvate concentration a sigmoidal velocity profile which is typical of an allosteric enzyme (Fig. 1). The $K0.5(PEP)$ values for this enzyme is estimated at 1.1mM PEP. However, in the presence of 0.5mM FDP the response is converted to a hyperbolic curve and the apparent Km is 0.15mM. With FDP, cooperativity was greatly reduced with Hill slopes of 0.9. Without FDP, the Hill slope was 2.8 (Fig. 2). Similar observations on FDP activation are made on pyruvate kinase isolated from *S. mansoni* miracidia.

Similar substrate saturation profiles were determined for pyruvate kinase from *S. mansoni* and similar results were obtained (Fig. 3). The $K0.5(PEP)$ value in the absence of FDP was 1.2mM PEP. In the presence of FDP the apparent Km is decreased to 0.11mM. Hill plots similar to those reported for *S. japonicum* were also obtained (Fig. 4). Again, with FDP, cooperativity was greatly reduced with Hill slopes of 0.84 and without FDP the Hill slopes of 0.84 and without FDP the Hill slope was 2.8. The apparent K_m for ADP was 1.72mM and 1.39mM for *S. japonicum* and *S. mansoni*, respectively. These values were apparently unaffected by the addition of FDP (Figs. 5,6). The pH optimum for the pyruvate kinase from both species of schistosome was 7.5 (Fig. 7). This value was unaffected by the addition of FDP.

As can be seen in Table 2, ATP is a potent inhibitor of schistosome pyruvate kinase. At low concentrations of Mg^{++} and 5mM ATP the inhibition can be as great as 99%. However, the data presented in Table 2 indicate that assessing the apparent inhibition by ATP is complicated by the relationship of Mg^{++}

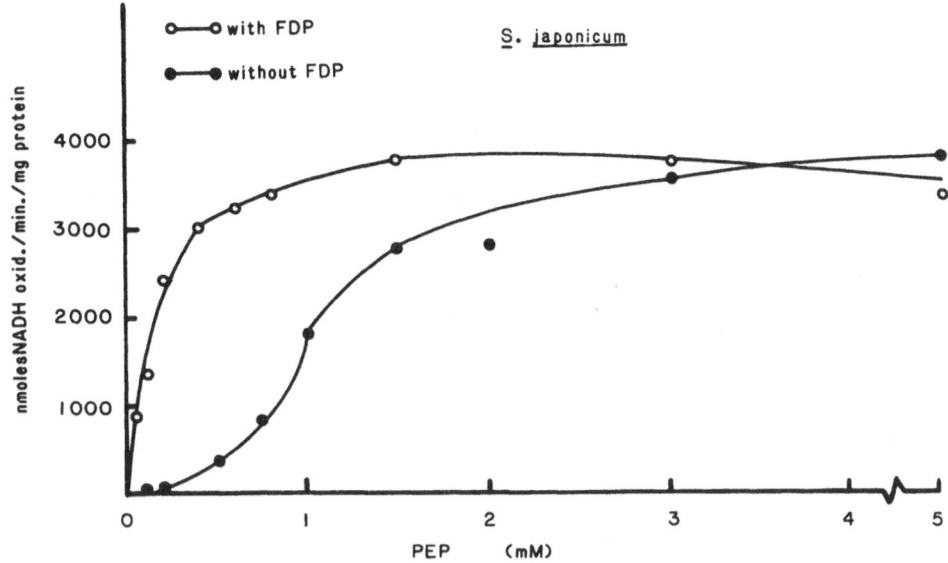

FIGURE 1. Substrate concentration-velocity curves for pyruvate
kinase in the 100,000xg supernatant obtained from
homogenates of adult *S. japonicum* paired worms. Assay
conditions are identical to those used above with *S.
mansoni*.

and ATP. This may simply indicate that the ATP effect is partially
due to the formation of $Mg^{++}ATP$ complexes as the inhibition is
partially overcome by increasing the Mg^{++} contration (Holmsen and
Storm, 1969). Ca^{++} also inhibits schistosome pyruvate kinase very
effectively as it does pyruvate kinase from other organisms
(Table 2). These data indicate that Ca^{++} inhibition is also
influenced by Mg^{++} concentration. However, no inhibition was
observed with alanine nor any amino acid known to inhibit
mammalian pyruvate kinase. This is consistent with observations
from other parasites and may well reflect a fermentative rather
than a gluconeogenic function of pyruvate kinase from these
parasites (Prichard and Schofield, 1968).

DEAE-chromatography of pyruvate kinase from both species
reveals a single peak eluting with 0.1M NaCl (data not shown). The
enzyme obtained from DEAE chromatography was concentrated and re-
chromatographed on Sephadex G-200. Pyruvate kinase activity was

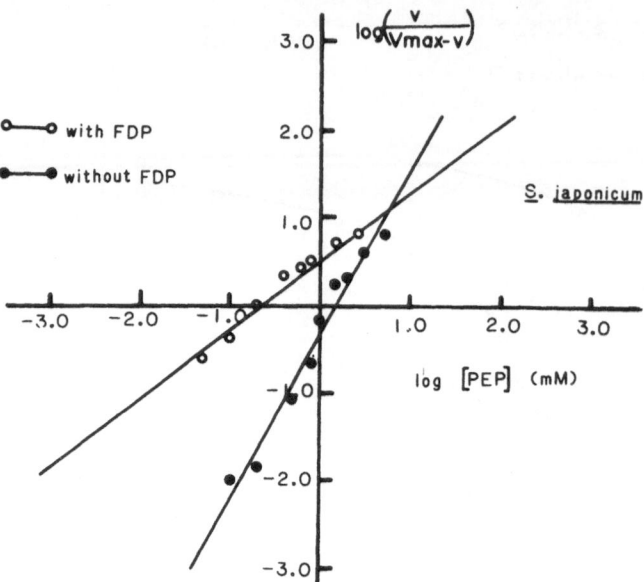

FIGURE 2.   Relationship between initial velocity of *S. japonicum* pyruvate kinase and total PEP concentration. Activity was determined as described in Fig. 1. Hill slopes are discussed in the text.

eluted from the column at a position consistent with an estimated molecular weight of 250,000.

From the data presented, it is clear that pyruvate kinase from *S. mansoni* and *S. japonicum* are well regulated enzymes. The apparent discrepancies in the regulatory parameters of *S. mansoni* pyruvate kinase observed by us and those observed by Brazier and Jaffe (1972) remain an enigma.

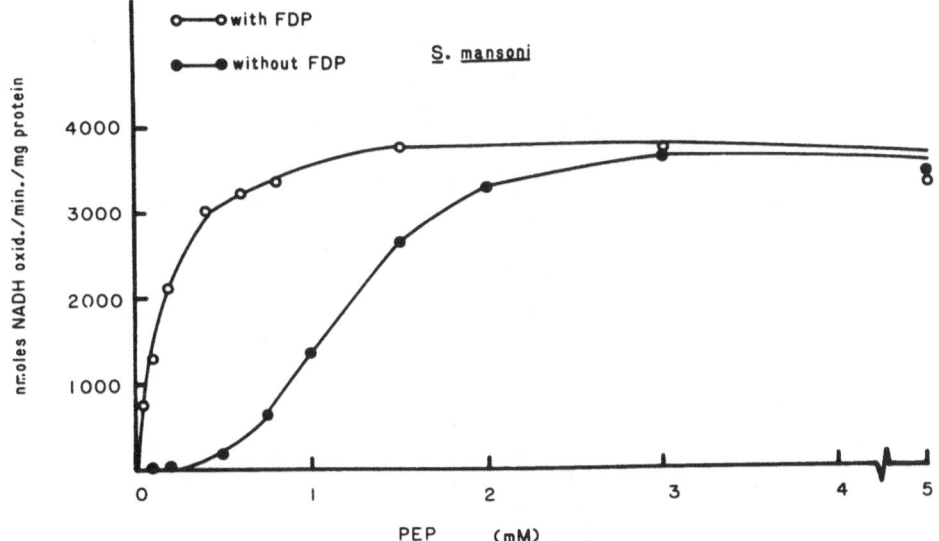

FIGURE 3. Substrate concentration-velocity curves for pyruvate
kinase in the 100,000xg supernatant obtained from
homogenates of adult *S. mansoni* paired worms. Closed
symbols: optimal concentrations of ADP (5mM), Mg^{++}
(5mM), K^+(40mM) NADH(0.2mM) in 100mM Imidazole buffer
at pH 7.5, and excess lactate dehydrogenase. Open
symbols: as above in the presence of 0.5mM fructose-
$1,6-P_2$.

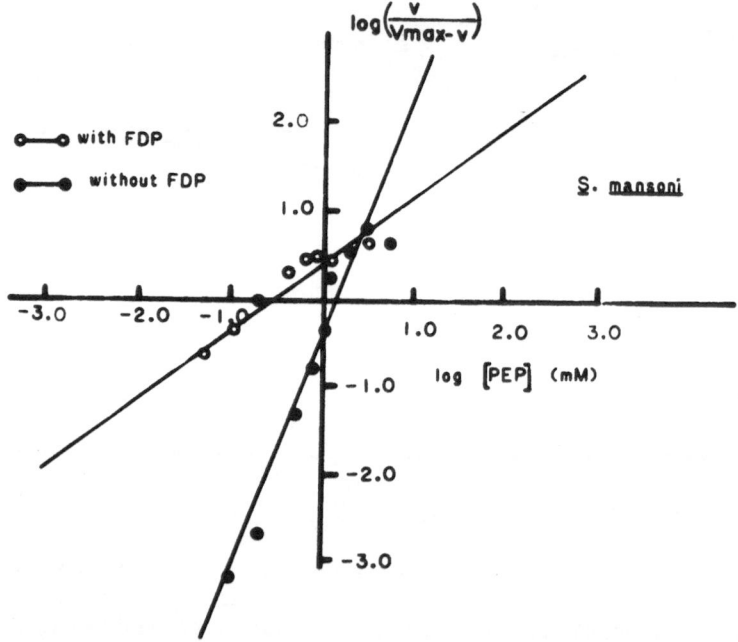

FIGURE 4.   Relationship between initial velocity of *S. mansoni*
            pyruvate kinase and total PEP concentration.   Activity
            was determined as described in Fig. 1.   Hill slopes are
            discussed in the text.

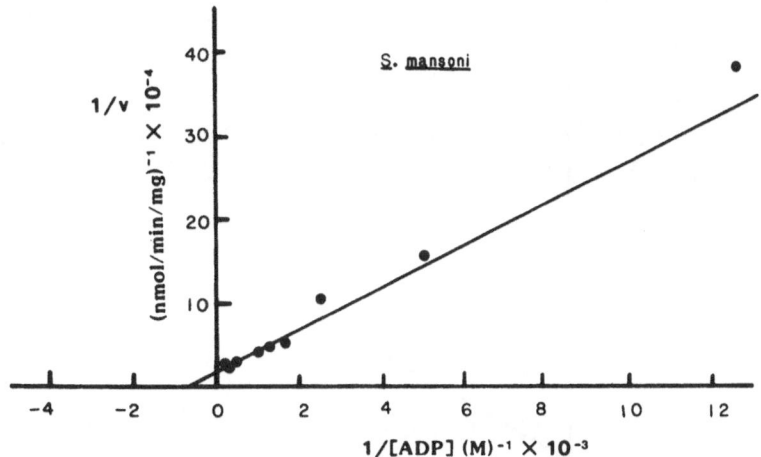

FIGURE 5.   Relationship between initial velocity of *S. mansoni*
pyruvate kinase (source in Fig. 1) and ADP concentration.   The assay mixture (1.0 ml) contained 100μmoles
Imidazole buffer at pH 7.5; 5μmoles Mg^{++}; 40μmoles
K^{+}; 5μmoles PEP, 0.2μmoles NADH; and excess lactate
dehydrogenase.   Closed symbols in the absence of FDP,
while open symbols represent the presence of 0.5μmoles
FDP.

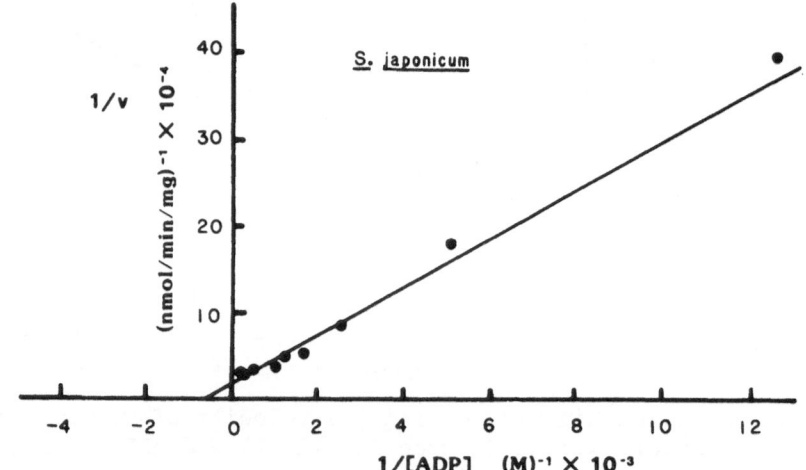

FIGURE 6.   Same conditions as Fig. 5 for *S. japonicum* pyruvate
             kinase.

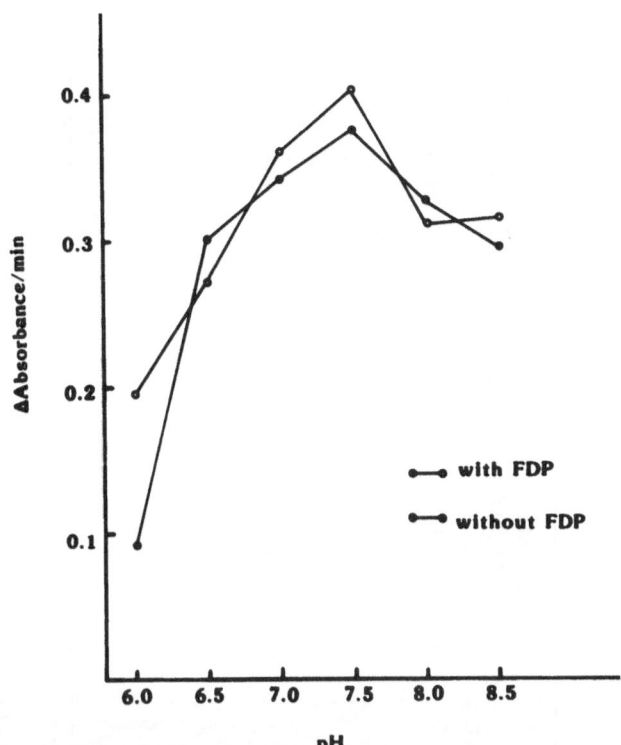

FIGURE 7.   pH-activity curves for pyruvate kinase in 100,000xg
           supernatant obtained from an homogenate of *S. mansoni*
           paired worms.   Activity determined under conditions
           described in Fig. 1.

TABLE 2. Inhibition of pyruvate kinase by ATP and CA^{++} in schistosomes.

ENZYME FRACTION	M_G^{++} (mM)	SPECIFIC ACTIVITY CONTROL*	% INHIBITION			
			ATP		Ca^{++}	
			1mM	5mM	1mM	5mM
S. mansoni	0.5	51.0	92.0	94.7	0	51.8
	2.5	1507.4	46.7	99.4	45.6	83.2
	5.0	2449.0	20.4	97.0	36.2	70.1
S. japonicum	0.5	98.0	87.7	92.9	0	55.4
	2.5	2770.0	60.6	99.7	49.6	85.5
	5.0	2942.0	4.2	97.9	13.2	76.6

*ng NADH oxid/min/mg Protein

II. REFERENCES

Barrett, J. and Beis, I. (1973). Nicotinamide and adenosine
    nucleotide levels in *Ascaris lumbricoides, Hymenolepis diminuta*
    and *Fasciola hepatica*. *Intl. J. Parasit.*, 3, 271-273.

Brazier, J. B. and Jaffe, J. J. (1972). Two types of pyruvate
    kinase in schistosomes and filariae. *Comp. Biochem. Physiol.*,
    44B, 145-155.

Bueding, E. (1950). Carbohydrate metabolism of *Schistosoma
    mansoni*. *J. Gen. Physiol.*, 33, 475-495.

Bueding, E. and Saz, H. J. (1968). Pyruvate kinase and phos-
    phoenolpyruvate carboxykinase activities of *Ascaris* muscle,
    Hymenolepis diminuta and *Schistosoma mansoni*. *Comp. Biochem.
    Physiol.*, 24, 511-518.

Carminatti, H., Jimenez de Asua, L., Liederman, B., and Rosengurt,
    E. (1971). Allosteric properties of skeletal muscle pyru-
    vate kinase. *J. Biol. Chem.*, 246, 7284-7288.

Carter, C. E. and Fairbairn, D. (1975). Multienzymic nature of
    pyruvate kinase during development of *Hymenolepis diminuta*.
    J. Exp. Zool., 2, 439-448.

Holmsen, H. and Storm, F. (1969). The adenosine triphosphate
    inhibition of the pyruvate kinase reaction and its dependence
    on the total magnesium ion concentration. *Biochem. J.*, 112,
    303-315.

Jacobson, K. W. and Black, J. A. (1971). Conformational differences
    in the active sites of muscle and erythrocyte pyruvate kinase.
    J. Biol. Chem., 246, 5504-5509.

Jimenez de Asua, L., Rozengurt, J. E., Devalle, J., and
    Carminatti, H. (1971). Some kinetic differences between the
    M-isoenzymes of pyruvate kinase from liver and muscle.
    Biochem. Biophys. Acta, 235, 326-334.

Llorente, P., Marco, R., and Sols, A. (1970). Regulation of liver
    pyruvate kinase and phosphoenolypyruvate crossroads. *Eur.
    J. Biochem.*, 13, 45-54.

Pritchard, R. K. and Schofield, P. J. (1968). The metabolism of
    phosphoenolpyruvate and pyruvate in the adult liver fluke
    Fasciola hepatica. *Biochem. Biophys. Acta*, 170, 63-76.

Saz, H. J. (1970). Comparative energy metabolisms of some
    parasitic helminths. *J. Parasitol.*, 56, 634-642.

Saz, H. J. (1971). Facultative anaerobiosis in the invertebrate.
    Pathways and control systems. *Am. Zool.*, 11, 125-135.

Tanaka, T., Sue, R., and Morimura, H. (1967). Feed forward
    activation and feedback inhibitors of pyruvate kinase type
    L of rat liver. *Biochem. Biophys. Res. Commun.*, 29, 444-449.

Vivayvargiya, R., Schwark, W. S., and Singhal, R. L. (1969).
    Pyruvate kinase: modulation by L-phenylalanine and L-alanine.
    Can. J. Biochem., 47, 895-898.

Watts, S. D. M. and Fairbairn, D. (1974). Anaerobic excretion
    of fermentation acids by *Hymenolepis diminuta* during
    development in the definitive host. *J. Parasitol.*, 60,
    621-625.

MORPHOLOGICAL AND HISTOCHEMICAL ADAPTATIONS OF TREMATODE DIGESTION,

WITH PARTICULAR EMPHASIS ON *Schistosoma mansoni*

Burton J. Bogitsh

Department of Biology
Vanderbilt University
Nashville, Tennessee   37235

The morphology of the digestive system of digenetic trematodes has been reviewed by Bogitsh (1975) and Erasmus (1977). In general, the digestive system is divisible into two regions, the foregut and the caecum. The function of the former region appears to be the ingestion and assimilation of exogenous foodstuffs. In this capacity, it is a highly adaptable area, often displaying myriad specializations that increase efficiency. These specializations may include secretory cells such as is observed in *Brachycoelium salamandrae* (see Bogitsh and Ryckman, 1982), sensory structures as are present in *Megalodiscus temperatus* (see Bogitsh, 1972), etc. The lining of the foregut is a continuation of the general tegumental surface of the adult worm with an abrupt and sharp demarcation, a septate desmosome, between it and the gastrodermis.

The gastrodermis characteristically displays two types of surface amplifications of the luminal plasma membrane. One type is a flattened leaf-like, or lamellate, type. Fujino and Ishii (1979) subdivided lamellar surface amplifications into three morphological categories: slender, ribbon-shaped; broad, triangular filamentous; broad, sheet-like or triangular. The second type of surface amplification is a digitiform, or microvillous, type. The gastrodermis itself may be either syncytial (*Schistosoma mansoni*) or cellular (*Haematoloechus medioplexus*).

One of the major drawbacks to the study of functional aspects of the digestive system of digeneans is the absence of a reliable *in vitro* system that would enable the investigator to manipulate feeding schedules with inhibitors, etc. Clegg and Smithers (1972) have perfected a penetration apparatus that produces schistosomules of *Schistosoma mansoni* from cercariae *in vitro*. A number of studies have evolved from the utilization of the *in vitro* transformation of the cercariae of *S. mansoni*, and, for this reason, there is probably more information available concerning the functional aspects of the schistosome digestive tract than any other digenean. The remaining discussion will, therefore, be limited to observations drawn from the study of these organisms.

II. REVIEW AND DISCUSSION

The report of Faust and Meleney (1924) is one of the earlier studies indicating the importance of red blood cells to the nutrition of schistosomes. Cheever and Weller (1958) studied the nutritional requirements of schistosomules recovered primarily from the livers of infected mice and subsequently maintained *in vitro*. Using increase in size as the index for determining the portion of host blood that possessed growth-stimulating properties, they reported that the stroma and the supernatant hydrolysate of mouse red blood cells retained a degree of growth-promoting activity although to a lesser extent than either whole red blood cells or a combination of stroma and hydrolysate. Their studies suggest that the stimulatory effect of red blood cells on the growth of the organisms *in vitro* is not due primarily to their hemoglobin content. Cheever and Weller's (1958) investigation is an extension of work of Senft and Weller (1956) who had reported previously that schistosomules recovered from infected mice grow *in vitro* only after red blood cells are added to the culture medium. Timms and Bueding (1959) reported the presence of a hemoglobin-specific protease in *S. mansoni* adults. Senft and his co-workers (see Senft, 1976 for review) further characterize this enzyme and elaborate on its ability to digest the globin portion of the molecule. One of the major drawbacks to the study of ingestion and digestion of food in schistosomes is the inability of the adult worms to feed *in vitro* (Zussman et al., 1970; Bogitsh, 1977). To circumvent this problem, Zussman et al. (1970) report that they labeled host hemoglobin with [^{3}H]-leucine via reticulocytes and, subsequently, fed the labeled cells to adult worms *in vivo*. Radioautographic studies show extensive distribution of the incorporated leucine throughout the tissues. The first to display the labeled amino acid is the gastrodermis, leading these investigators to conclude that the amino acid is derived from the enzymatic breakdown of globin and transported from the lumen into the gastrodermis.

Clegg and Smithers (1972) demonstrate that feeding habits and growth patterns of *S. mansoni* schistosomules, collected after skin penetration and maintained *in vitro*, closely parallel *in vivo* patterns. Their technique provides a distinct advantage in that, since the organisms are not exposed to host blood, it is possible to manipulate foodstuffs and schedules without contamination by host blood. Using this procedure, Bogitsh and Carter (1977) have studied the development of the digestive tract of *S. mansoni* schistosomules during the first 15 days post-penetration, including the development of the esophageal gland and the effect of feeding on its development.

Bogitsh (1977) has utilized drugs such as colchicine and vin-blastine to demonstrate the sensitivity of schistosomules to these alkaloids relative to feeding and secretory activities. Using cytochemical techniques, Bogitsh (1978) also has confirmed the ability of the schistosomules to ingest exogenous hemeproteins *in vitro* and has charted the fate of these molecules in the digestive tract lumen. These studies show that newly penetrated schistosomules are capable of ingesting hemeproteins in solution at a very early stage although the ability to ingest red blood cells is not acquired until 4 to 6 days post-penetration (Bogitsh and Carter, 1977). Once the red blood cell is ingested, its structural integrity is lost in the distal portion of the foregut, with the final stages of digestion occurring in the caecum.

Investigations have also shown at least two types of secretory granules associated with the digestive tract of schistosomes. One, a membranous granule, is produced and secreted from the esopha-geal gland (Morris and Threadgold, 1968; Spence and Silk, 1970; Dike, 1971; Bogitsh and Shannon, 1971; Ernst, 1975). It appears likely that the secretion or translocation of the granule is dependent on microtubules, considering the effects of both colchicine and vinblastine (Bogitsh, 1977).

A second granule is produced in the gastrodermis (Morris, 1968). The translocation of this granule is apparently unaffected by these microtubule-depolymerzing drugs (Bogitsh, 1977). From these observations, it seems reasonable to hypothesize that the two granules are involved in the lysis and digestion of red blood cells in the lumen of the digestive tract of the schistosomes. Data also indicate the possibility that esophageal granules are responsible for the initial lysis of red blood cells in the lumen of the esophagus and that the gastrodermal secretions may produce the specific "hemoglobinase" complex responsible for the digestion of the globin portion of the molecule in the lumen of the caecum (Bogitsh, 1981). This hypothesis seems particularly tenable since

the proximity of the absorptive surface to the site of digestion
affords the worm a spatial advantage in having both processes
occurring in the caecal area. The hypothesis is strengthened further
by the fact that schistosomules treated with either colchicine or
Actinomycin-D have a demonstrable ability to ingest dissolved hemoglobin; however, of the two, only the colchicine-treated worms
possess the ability to digest the hemeprotein (Bogitsh, 1981).

Despite the fact that adult schistosomes reside in an environment rich in free amino acids, there is little doubt that growth
and development of the worms are dependent upon ingestion and
digestion of host hemoglobin (Senft and Weller, 1956; Cheever and
Weller, 1958; Clegg and Smithers, 1972; Bogitsh and Carter, 1977).
The quest for the isolation, characterization, and site of formation of the requisite digestive enzyme(s) has gained impetus from
the report of Timms and Bueding (1959) who studied a proteolytic
enzyme from homogenized adult worms. The enzyme has a pH optimum
of 3.9 and a marked substrate specificity for hemoglobin. They
hypothesize that the enzyme is active in the intestine of the
schistosomes. Senft and his associated (Grant and Senft, 1971;
Sauer and Senft, 1972; Senft, 1976), Deelder et al. (1977), and
Dresden and Deelder (1979) also have isolated an acid proteinase
from adult schistosomes. The latter investigators further report
that the acid proteinase resembles lysosomal cathepsin B in a
number of ways. For example, both are sensitive to the same
inhibitors (e.g., leupeptin, dibromoacetophenone, N-ethylmaleimide)
and activators (e.g., thiols), and both are active against the
same natural substrates (e.g., hemoglobin and trypsinogen).

The studies of Maki et al. (1982) deal with acid protease
activity in a variety of parasitic helminths, including a number
of blood-feeding nematodes. In the forms investigated, they
report that the ability to digest hemoglobin is due, at least in
part, to an acid carboxyl proteinase similar to cathepsin D rather
than to a thiol proteinase as described for S. mansoni.

Sauer and Senft (1972) and Senft (1976) further report that
the "hemoglobinase" activity of the isolated S. mansoni proteinase
produces only peptides of 8-10 amino acids in length and very few
amino acids from hemoglobin. A similar phenomenon is reported
by Simpkin et al. (1980) regarding a semipurified proteinase from
the gut exudate of Fasciola hepatica. The F. hepatica enzyme also
displays a marked affinity for globin with a pH optimum of 3.9-4.0.
This enzyme incompletely hydrolyzes globin, producing polypeptides
and a few free amino acids. Since it is unlikely that polypeptides
larger than dipeptides can diffuse through the gastrodermal membrane
(Ehrenreich and Cohn, 1969), it is clear that diffusible molecules
must be produced by some other means. The characteristics of the
gastrodermis of schistosomes, and digenetic trematodes in general,

make it doubtful that phagocytosis or pinocytosis occurs; hence
endocytosis is precluded as a vehicle for the entry of macromolecules into the tissue (Bogitsh, 1975; Erasmus, 1977). It is
logical to assume, therefore, that a mechanism exists, as in
vertebrates (Coffey and DeDuve, 1968), for degradation of globin
into diffusible components. The action of other acid proteases,
such as dipeptidyl peptidases I and II, may be required to supple-
ment the activity of a cathepsin B-like enzyme in order for polypeptides to be degraded into diffusible dipeptides. The following
is suggested as a possible mechanism:

Globin $\xrightarrow{\text{"hemoglobinase" (cathepsin B)}}$ Polypeptides $\xrightarrow{\text{DAP I AND II}}$

                              Dipeptides.

     Grant and Senft (1971), in discussing the possible physiological
role of the "hemoglobinase" and the consequence of incomplete
hydrolysis of globin, interpret Fripp's (1967) report of the lack
of gastrodermal leucine aminopeptidase activity in *Schistosoma
rodhaini* as indicative of a total lack of peptidase activity. This
interpretation fails to recognize that peptidases other than
aminopeptidases may be present. Bogitsh and Dresden (1983), report
that, while leucine and alanine aminopeptidase activity cannot be
demonstrated histochemically in *S. mansoni* and *Schistosoma japonicum,*
both DAP I and DAP II are present in the gastrodermis. Reaction
product for DAP I and DAP II, as well as for cathepsin B, is
observed as discrete granules in the gastrodermis. No activity is
detected in the foregut or esophageal glands.

     In light of previous biochemical studies on "hemoglobinase"
activity in the schistosomes, it appears likely that a battery of
acid proteases is involved in the ultimate digestion of host
hemoglobin. The lack of acid protease activity in the esophageal
glands and in the foregut is consistent with the theory that the
esophageal gland secretions are not responsible for the digestion
of host hemoglobin.

     Additional studies provide a cytochemical and morphological
profile of the gastrodermal Golgi complexes and help to clarify
the role of the organelle (Bogitsh, 1981). These studies will
take on added significance is a definite relationship can be
established between the Golgi complex and a "hemoglobinase"-
containing secretory granule. Bogitsh (1982) reports that the
gastrodermal Golgi complex demonstrates a functional bipolarity
with the outer cisternae involved in the packaging and emission of
one type of vesicle, perhaps a lysosome, and the inner cisternae
concerned with the formation of the larger, membrane-bound,

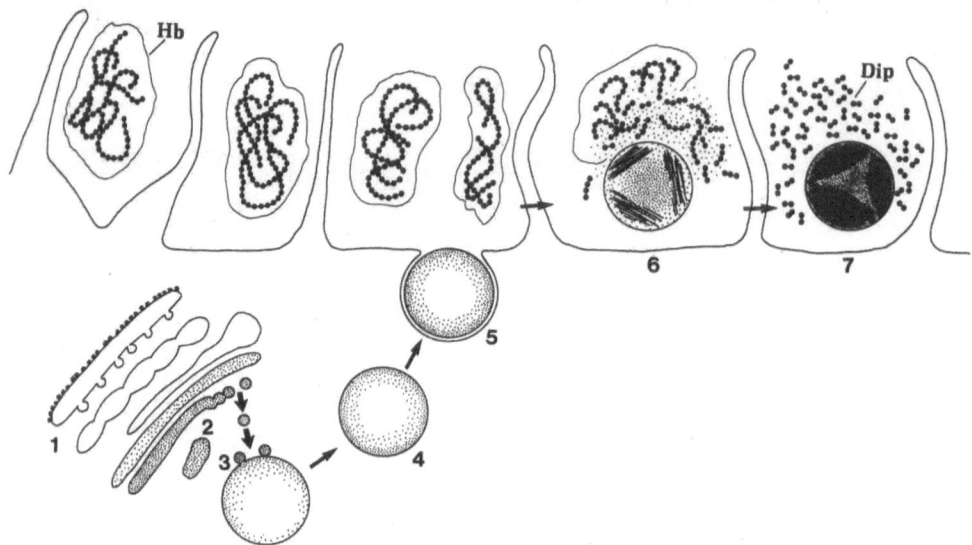

FIGURE 1.   Diagram of proposed pathway of hemoglobin (Hb) digestion
            by schistosomes and interrelationship between various
            organelles in this process. Proteases are synthesized
            in the Golgi apparatus (1), packaged in secretory
            secretory vesicles (2) and released at the trans face of
            the Golgi (2). The secretory vesicles fuse with lipid-
            like droplets (3) which, in turn, migrate to luminal
            surface of gastrodermis (4) where they are expelled into
            lamellate-restricted areas (5) where hemoglobin (Hb) is
            concentrated. In contact with substrate Hb, proteases
            interact with substrate and initiate sequential enzyme-
            substrate interactions (6) which will hydrolyze Hb mole-
            cule eventually into diffusible dipeptides (Dip) (7).
            Lipid -like droplets become increasingly electron-dense
            (5,6,7) as enzymes are released into lumen and are
            eventually regurgitated as part of "schistosome pigment."

secretory granule. As mentioned above, the translocation of these granules is apparently unaffected by colchicine or vinblastine.

Hemoglobin digestion occurs in the lumen of the caeca close to the absorptive suface of the gastrodermis (Bogitsh, 1981). Davis et al. (1969) hypothesize that digestion of blood in the blood-feeding trematode *Haematoloechus medioplexus* occurs in an area delimited by surface amplifications which they term "superficial digestive vacuoles." They further suggest that the lipid-like globules observed in the gastrodermis, rich in lysosomal enzymes, empty their contents into these restricted areas causing food to be digested as close to the absorptive surface as possible. It is possible that a similar mechanism may function in the schistosomes (Fig. 1).

As indicated above, almost all research on the use of hemo-globin by schistosomes has been centered upon the globin moiety. It has long been considered that the iron porphyrin moiety of the molecule is regurgitated. In an attempt to determine whether or not schistosomules of *S. mansoni* can incorporate precursors of hemoglobin synthesis other than amino acids, labeled δ-amino-levulinic acid (ALA) was added to the *in vitro* culture medium used for the maintenance of the larvae. Subsequent radioautographs demonstrated incorporation of label throughout the tissues of the schistosomules. Further investigation of this phenomenon required the labeling of the porphyrin moiety of the hemoglobin molecule via the incorporation of either [3H]-or [14C]-labeled ALA in mouse reticulocytes. Following the labeling procedure, the re-sulting labeled red blood cells were fed to schistosomules *in vitro*. After approximately 5 days of feeding, the larvae were wash-ed, pooled, homogenized, and centrifuged. The resulting pellets were subject to trichloracetic acid (TCA) precipitation and analyzed for label. The results of this experiment indicate that since a significant number of counts are found in TCA precipitable fractions, the worms are not only capable of using portions of the porphyrin moiety, but they can incorporate it into macromolecules that are probably protein.

III. REFERENCES

Bogitsh, B. J. (1972). Cytochemical and biochemical observations on the digestive hosts of digenetic trematodes. IX. *Megalodiscus temperatus. Exp. Parasitol.*, 32, 244-266.

Bogitsh, B. J.   (1975).   Cytochemical observations on the gastro-
    dermis of digenetic trematodes. *Trans. Amer. Micros. Soc.*,
    94, 524-528.

Bogitsh, B. J.   (1977). *Schistosoma mansoni:* colchicine and
    vinblastine effects on schistosomule digestive tract develop-
    ment *in vitro*. *Exp. Parasitol.*, 43, 180-188.

Bogitsh, B. J.   (1978). *Schistosoma mansoni:* uptake of exogenous
    hemeproteins by schistosomules grown *in vitro*. *Exp.
    Parasitol.*, 45, 247-254.

Bogitsh, B. J.   (1981). *In vitro* effects of inhibitors on ingestion
    and digestion of hemoglobin by *Schistosoma mansoni*
    schistosomules. *J. Parasitol.*, 67, 875-880.

Bogitsh, B. J.   (1982). *Schistosoma mansoni:* cytochemistry and
    morphology of the gastrodermal Golgi apparatus. *Exp.
    Parasitol.*, 53, 57-67.

Bogitsh, B. J. and Carter, O. S.   (1977).   Developmental studies
    on the digestive tract of schistosomules (*Schistosoma mansoni*)
    grown *in vitro*. I. Ultrastructure. *Trans. Amer. Micros.
    Soc.*, 96, 219-227.

Bogitsh, B. J. and Dresden, M. H.   (1983).   Fluorescent histo-
    chemistry of acid proteases in adult *Schistosoma mansoni*
    and *Schistosoma japonicum*. *J. Parasitol.*, 69, 106-110.

Bogitsh, B. J. and Ryckman, C. S.   (1982).   The ultrastructure of
    Brachycoelium salamandrae gastrodermis with observations on
    the effects of starvation. *J. Parasitol.*, 68, 824-833.

Bogitsh, B. J. and Shannon, W. A.   (1971).   Cytochemical and bio-
    chemical observations on the digestion tracts of digenetic
    trematodes. VIII. Acid phosphatase activity in *Schistosoma
    mansoni* and *Schistosomatium douthitti*. *Exp. Parasitol.*,
    29, 337-347.

Cheever, A. W. and Weller, T. H.   (1958).   Observations on the
    growth and nutritional requirements of *Schistosoma mansoni*
    in vitro. *Am. J. Hyg.*, 68, 322-339.

Clegg, J. A. and Smithers, S. R.   (1972).   The effects of immune
    rhesus monkey serum on schistosomules of *Schistosoma mansoni*
    during cultivation *in vitro*. *Intl. J. Parasitol.*, 23, 79-98.

Coffey, J. W. and DeDuve, C. (1968). Digestive ability of lyso-
    somes. I. The digestion of proteins by extracts of rat
    liver lysosomes. *J. Biol. Chem.*, 243, 3255-3263.

Davis, D. A., Bogitsh, B. J., and Nunnally, D. A. (1969). Cyto-
    chemical and biochemical observations on the digestive tracts
    of digenetic trematodes. III. Nonspecific esterase in
    Haematoloechus medioplexus. *Exp. Parasitol.*, 24, 121-129.

Deelder, A. M., Reinders, P. N., and Rotmans, J. P. (1977).
    Purification studies on an acidic protease from adult
    Schistosoma mansoni. *Act. Leiden.*, 45, 91-103.

Dike, S. C. (1971). Ultrastructure of the esophageal region in
    Schistosoma mansoni. *Amer. J. Trop. Hyg.*, *20*, *552-568*.

Dresden, M. H. and Deelder, A. M. (1979). *Schistosoma mansoni:*
    thiol proteinase porperties of adult worm "hemoglobinase."
    Exp. Parasitol. 48, 190-197.

Ehrenreich, B. A. and Cohn, Z. A. (1969). The fate of peptides
    pinocytosed by macrophages *in vitro*. *J. Exp. Med.*, 129, 227-
    243.

Erasmus, D. A. (1977). The host-parasite interface of trematodes.
    Adv. Parasitol., 15, 201-242.

Ernst, S. C. (1975). Biochemical and cytochemical studies of
    digestive-absorptive functions of esophagus, cecum, and
    tegument in *Schistosoma mansoni:* acid phosphatase and tracer
    studies. *J. Parasitol.*, 61, 633-647.

Faust, E. C. and Meleney, H. E. (1924). Studies on schistosomiasis
    japonica. *Am. J. Hyg.* Monograph Series No. 3.

Fripp, P. J. (1967). The histochemical localization of leucine
    aminopeptidase in *Schistosoma rodhaini*. *Comp. Biochem. Physiol.*,
    20, 307-309.

Fujino, T. and Ishii, Y. (1979). Comparative ultrastructural topo-
    graphy of the gut epithelia of some trematodes. *Intl. J.
    Parasitol.*, 9, 435-448.

Grant, C. T. and Senft, A. W. (1971). Schistosome proteolytic
    enzyme. *Comp. Biochem. Physl.*, 38B, 663-678.

Maki, J., Furuhashi, A., and Yanagisawa, T. (1982). The
    activity of acid proteases hydrolysing haemoglobin in
    parasitic helminths with special reference to interspecific
    and intraspecific distribution. *Parasitology*, 84, 137–147.

Morris, G. P. (1968). Fine structure of the gut epithelium of
    Schistosoma mansoni. *Experientia*, 24, 480–487.

Morris, G. P. and Threadgold, L. T. (1968). Ultrastructure of
    the tegument of adult *Schistosoma mansoni*. *J. Parasitol.*,
    54, 15–27.

Sauer, M. C. V. and Senft, A. W. (1972). Properties of a proteo-
    lytic enzyme from *Schistosoma mansoni*. *Comp. Biochem.
    Physiol.*, 42B, 205–220.

Senft, A. W. (1976). Observations on the physiology of the gut
    of *Schistosoma mansoni*. *In* "Biochemistry of Parasites and
    Host-Parasite Relationships." (H. Van den Bossche, ed.),
    pp. 335–342. North Holland, Amsterdam/New York.

Senft, A. W. and Weller, T. H. (1956). Growth and regeneration
    of *Schistosoma mansoni in vitro*. *Proc. Soc. Exp. Biol. Med.*,
    93, 10–19.

Simpkin, K. G., Chapman, C. R., and Coels, G. C. (1980). *Fasciola
    hepatica*: a proteolytic digestive enzyme. *Exp. Parasitol.*,
    49, 281–287.

Spence, I. M. and Silk, M. H. (1970). Ultrastructure studies of
    the blood fluke – *Schistosoma mansoni*. IV. The digestive
    system. *S. Afr. J. Med. Sci.*, 70, 93–112.

Timms, A. R. and Bueding E. (1959). Studies of a proteolytic
    enzyme from *Schistosoma mansoni*. *Brit. J. Pharm. Chemo.*,
    14, 68–73.

Zussman, R. A., Bauman, P. M., and Petruska, J. C. (1970). The
    role of ingested hemoglobin in the nutrition of *Schistosoma
    mansoni*. *J. Parasit.*, 56, 75–79.

PARASITE IMMUNE ESCAPE MECHANISMS

W. M. Kemp

Department of Biology
Texas A&M University
College Station, Texas   77843

I. INTRODUCTION

In contemplating how best to present this topic I was faced with two approaches. One was to catalogue all the parasites about which we know something of immune escape and list the mechanisms associated with each. The other approach was to choose one parasite and to detail our knowledge of escape mechanisms in a more exhaustive way. I have elected to pursue the second approach so that the subject may be examined in depth and the complexities of the mechanisms more easily discerned.

The parasite I have chosen is *Schistosoma mansoni*. The reasons
for this choice are: (1) The parasite has a complex life cycle with
an equally complex relationship with its host. (2) This is one of
the best studied parasite systems in regard to immune escape.
(3) Schistosomes are among the medically most important parasites
in the world. (4) This host-parasite relationship is the one with
which I am most familiar and upon which my laboratory has concentrated its efforts. What follows is a highly personalized view of
the field of schistosome immune escape and, as such, may or may
not be an accurate reflection of the opinion of other workers in
this discipline.

Parasitologists have acknowledged for some time that successful
parasites are those which have interdigitated themselves with their
hosts at the molecular level. In most chronic parasitic infections
a precarious balance must be achieved between establishing and
maintaining a successfully reproducing population of parasites while
avoiding overpopulation and excessive pathology so that early death
of the host does not destroy the parasite's habitat. The schistosomes have responded to this problem by evolving a relationship with
the definitive host in which the adult parasites survive and reproduce for decades, while newly penetrating and migrating larvae of
subsequent infections are detected and destroyed by the host's
immune response. This circumstance was first suggested by the work
of Smither et al. (1969) and has been termed concomitant immunity.

The life cycle of the schistosomes is less complex than most
trematodes in that the necessity for a second intermediate host, or
some other agent of metacercarial transmission, has been eliminated.
The schistosomes are also unusual among trematodes in that the
sexes are seperate and they inhabit the circulatory system of the
host, a site usually considered one of the most immunologically
hostile. From this niche it is impossible for the parasites to
successfully transmit their eggs to the external environment without inducing pathology. Eggs laid in the mesenteric venules must
lyse their way through the intestinal walls to gain access to the
lumen for elimination. Many of the eggs are swept by the venous
flow to the liver where they are sequestered and walled off by a
granulomatous response. Those eggs which do escape the host via the
feces hatch in water releasing a free-swimming larval stage, the
miracidium, which must locate and penetrate an appropriate molluscan host. The successful miracidium undergoes a developmental
transformation to a mother sporocyst stage which in turn produces
daughter sporocysts. Each daughter sporocyst produces numerous
cercariae which migrate out of the snail, swim free in the water,
and seek the definitive host. Upon contact with an appropriate
host, the cercaria penetrates the skin by means of a set of
highly developed penetration gland cells and commences a transformation into a new developmental stage called a schistosomule.

This transformation is both physiologically and anatomically dramatic.

The parasite change from a water adapted cercaria to a saline adapted schistosomule in a matter of a few minutes. The cercarial tail is lost, as is a micron thick glycocalyx which had totally encompassed the cercarial body.

The schistosomule now begins an odessey of morphological and physiological change as well as a change of location within the host. Although the term "schistosomule" is used to describe the migratory parasite from the time its cercarial characteristics are lost until the parasite displays the anatomical qualities of adulthood in the liver, the parasite itself is undergoing profound changes and appears to possess different biological characteristics at various points in the migration process. These changes are reflected in the escape mechanisms manifested during this process.

My comments regarding schistosome immune escape mechanisms will be limited to those stages found in the definitive host only. There is growing evidence of escape mechanisms in the snail host, but these mechanisms and the interplay between them and the molluscan defense system are sufficiently different from the mammalian host-parasite relationship to justify considering them in an alternate category deserving of in depth treatment in their own right.

II. LARVAL ESCAPE MECHANISMS

As noted in the previous discussion of concommitant immunity, there is a considerable difference between the obstacles faced by primary and secondary infection larvae. Approximately 65% of primary infection schistosomula die at some point during migration for reasons as yet unknown (possibly defects within the parasite itself) while less than 5% of secondary infection schistosomula survive to adulthood. From these observations it is obvious that larval escape mechanisms are for the most part limited to the primary infection when the host is unprepared to respond. These mechanisms are probably as transitory in nature as the larvae themselves and good for one time and one time only. They function to spare the initial parasite infection during the arduous and dangerous migration and maturation phases and are less effective after a primary infection has been established and the host is sensitized.

III. CERCARIA TO SCHISTOSOMULE TRANSITION

As previously outlined, a remarkable and profound series of changes occur when cercariae penetrate the host and begin to transform into schistosomula. The sum total of these changes

may add up to an "antigenic smokescreen" behind which the newly
formed schistosomula might escape detection and targeting by the
host's defenses. Cercarial stage specific antigens, which are lost
during transformation and never reexpressed by the later developmental stages, would form the main substance of this smokescreen.
Cercarial antigens associated with the penetration gland cells,
glycocalyx, and tail (which is sometimes discarded in the dermis
rather than outside the host) would fall into this category. The
glycocalyx has also been shown to carry snail antigens borrowed
from the intermediate host (Capron et al., 1968 Kemp et al., 1974).
The release of these molluscan antigens with the glycocalyx in
the host might serve to convince the host's immune response to
mount an attack on a presumed snail rather than a trematode invader.

The combined effect of these sources of superfluous antigenic stimulation could occupy a considerable portion of the
potential host immune response, while the parasite has time to
accomodate itself to its new environment and begin its migration
to the lungs.

One point of interest in this regard has to do with the chemical
nature of part of the glycocalyx. Normal definitive host serum has
been shown to kill cercaria by means of alternative complement pathway activation. Immune definitive host serum does not kill cercaria
in spite of the fact that antibodies bind to the glycocalyx and
classical complement pathway activation occurs (Stirewalt, 1963;
Kemp, 1970, 1972). Apparently antibodies bind to the glycocalyx
molecules responsible for alternative pathway activation and
block the activation process. This complement fixing characteristic
is lost as the glycocalyx is lost during transformation thus obviat-
ing the danger to the parasite of being killed by the alternative
pathway after penetration (Samuelson et al., 1980). We will return
to this observation in discussing adult worm immune escape.

Furthermore, at some time during this period a profound change
occurs at the tegumental surface and this change is manifested in
the conversion of the cercarial trilaminate tegument membrane to
a pentalaminate membrane which will be characteristic of all further
developmental stages through the adult (Hockley and McLaren, 1973;
McLaren et al., 1975, 1978). The origin, nature, and function of
this membrane change is currently under intensive study, but the
role of the membrane in the host-parasite relationship has yet to
be determined.

IV. PRE-LUNG STAGE SCHISTOSOMULA

Primary infection schistosomula in route from the skin to
the lungs may repeat some of the apparently successful ploys used
during the transition phase. However, at this time the parasite

first displays certain abilities which will be characteristic
throughout the remaining larval stages and adulthood. Primary
among these characteristics is the ability to acquire and bind
host antigens. Smithers et al. (1969) were the first to suggest
that adsorbed host antigens might function in aiding parasite
survival. A host antigen coated parasite would present fewer antigens to the host for a response and these adsorbed host molecules
could also sterically hinder antibodies or effector cells from
binding to their respective antigens, a mechanism which could
insure survival of the parasite even in the presence of a potentially
protective immune response.

Among the host antigens shown to be bound to the pre-lung
schistosomular surface are erthrocyte antigens (Clegg et al., 1971;
Dean, 1974), Forssman antigens (Dean and Sell, 1972), histocompatibility antigens (Sher et al., 1978), IgG-Fc receptors (Torpier
et al., 1979), and receptors for Clq and C3b (Santoro et al.,
1979; Ouaissi et al., 1980). In an associated activity the parasite seems to be able to enzymatically cleave the Fab portions of
Fc receptor bound IgG antibodies, leaving only the Fc fragment on
the tegumental surface. Furthermore, the polypeptides generated
from this cleavage appear capable of modulating certain immune
responses (Auriault et al., 1980, 1981). The physiological controls
of such a process are obviously sophisticated and highly adapted
to the definitive host.

V.   LUNG STAGE SCHISTOSOMULA

Lung stage schistosomula are somewhat of an anomaly. Reports
are inconsistant regarding the presence or absence of adsorbed host
antigens on the parasite's tegumental surfaces. Our own experience
in assaying for tegument-associated host immunoglobulin is that some
schistosomula have them and some do not. Whether this is a reflection of a developmental stage change (many helminth parasites
undergo significant changes in the lung environment), the anatomical
site of the parasite at the time of dissection (perhaps the parasite
was not in the circulatory system at the time), or of the sex of
the schistosomula (adult female worms do not adsorb host immunoglobulins) is not known at this time.

Of more significance were the observations of Dean (1977) that
lung stage parasites were totally refractory to immune attack.
This refractoriness was independent of host antigens and physiological processes. These studies have been confirmed by several
laboratories (Tavares et al., 1978, 1980; Samuelson et al., 1980;
Dessein et al., 1981; McLaren and Incani, 1982) and the additional
observation was made that these schistosomula remain refractory
to damage even if antibody is able to bind to the surface. Moser
et al. (1980) actually haptenated the schistosomular surface and

exposed the parasite to anti-hapten antibody plus complement, without
apparent deleterious effect. This refractory state is a temporary
one. Upon leaving the lungs in its continuing migration to the
liver the parasite once again becomes susceptible to antibody
mediated killing.

VI. POST-LUNG SCHISTOSOMULA

The post-lung schistosomule is similar to the pre-lung
schistosomule in almost every regard. The route of migration
from the lungs to the liver is a matter of considerable conjecture
and it may be that different parasites follow different routes.
Upon reaching the liver, the parasite, which has been growing
slightly during the migration phase, undergoes a dramatic growth
spurt. At some as yet undefined point in this maturation phase,
the parasite surface becomes the dynamic structure so characteristic
of the adult parasite.

VII. ADULT WORMS

The adult schistosome is the life cycle stage which has
received the most attention in terms of investigations into
mechanisms of immune escape, since it is this stage which survives
and flourishes in the immunologically primed host. There is
evidence that the host does eventually influence the parasite in
terms of size reduction (Ritchie et al., 1967; Cornford et al.,
1982) and reduced fecundity (Damian et al., 1976). The effects are
reversed when these worms are transplanted into an immunologically
naive host (Webbe, 1976).

Wilson and Barnes (1974, 1977) demonstrated a normal metabolic
turnover rate of the adult tegument of about 2-3 hr. This turnover
is apparently accompanied by a lateral shifting of membrane
associated substance to the tips of tegumental spines where dissociation from the parasites occurs.

The dynamics of the adult tegument when under stress was
illustrated by the discovery that the parasite has the ability to
rid itself of immunologically compromised surface antigens within
20 min (Kemp et al., 1980). This process, referred to as tegument
modulation, is a specific (only the compromised antigens are shed,
noncomprised antigens remain in place), energy dependent, micro-
filament mediated event. The presence of such a mechanism
broadens our appreciation of the tegument as a dynamic and resource-
ful interface with the host.

One of the most stimulating ideas of immune escape is the
concept of molecular mimicry (Damian, 1964, 1967). The original
theory proposed that parasites escape host immune responses by

genetically mimicking the biochemistry of the host, thereby
reducing the parasite's foreignness and its chances of recognition
and rejection. Damian argued that parasites exposed to the
selective pressures of the immune response would be expected to
ultimately become either more host-like or extinct. Proof of
antigen mimicry was presented by Damian et al. (1973) and confirmed
by Kemp et al. (1976a) with the demonstration of a murine antigen,
alpha 2-macroglobulin, on the surfaces of worms procured from
rhesus monkeys. Since primate and murine alpha 2-macroblobulins do
not immunologically cross-react, the only reasonable origin for
the murine antigen in the primate host is the parasite's genome.
Evidence for mimicry of molluscan antigens by the parasite was
presented by Capron et al. (1968) and Kemp et al. (1974).

The concept of adsorbed host components serving to camouflage
the larval parasite and block host effector mechanisms referred
to earlier (Smithers and Terry, 1967) was actually first applied
to the adult parasite. Considerable evidence has been presented
for this idea as it relates to adult escape. Some antigens which
have been documented to be associated with the adult worm are:
Erythrocyte antigen (Smithers et al., 1969), histocompatibility
antigens (Gitter and Damian, 1982), immunoglobulin (Sogandares-
Bernal, 1976; Kemp et al., 1976b; 1978), and the third component of
complement (Kabil, 1976; Rasmussen and Kemp, in press). Of particular interest in regard to tegument associated immunoglobulins
is that all classes and subclasses are present except IgD and IgE.
Furthermore, both heterospecific (antibody specific for antigens
other than the parasite) and homospecific (antibody specific for the
parasite) immunoglobulin have been shown to be associated with
the tegument by binding to Fc receptors in the membrane (Kemp
et al., 1977; Tarleton and Kemp, 1981; Rasmussen and Kemp, in
press). In addition, receptors for activated C3 have been shown
to be present on the parasite's surface (McGuinness and Kemp,
1981; Tarleton and Kemp, 1981).

The origin of these receptors is subject to question at the
present time. If they arise as expressions of the parasite's
genome they might serve as an interesting interface between the
concepts of antigen mimicry and antigen adsorption. If they
are borrowed from the host they still represent more than just a
simple adsorption of host material because they are intimately
associated with the parasite's cytoplasmic control mechanisms.
Recent work by Caulfield et al. (1980) has suggested a mechanism
by which schistosomula (and perhaps adult worms) could obtain
such antigens. Human neuptrophils were shown to bind to the
parasites without apparent damage to the worm, and leave behind
a portion of their membrane upon dissociation from the parasite.
Such an action could account for the appearance of the pentalaminate membrane, host antigens, and receptors for immunoglobulins

and C3. Caulfield has proposed that the membrane deposition would occur by a flowing mechanism so that surface components oriented toward the outside on the neutrophil would retain that orientation on the worm. The only inconsistancy between this concept and current data lies on the report of Gitter et al. (1982) that although major histocompatibility antigens are present on the adult worms, their orientation is such that they failed to stimulate a mixed lympho- cyte reaction. These data notwithstanding, the concept of Caulfield and his colleagues is fascinating and certainly worthy of further study.

Investigation continues into the possible roles of the Fc and C3 receptors in adult parasite escape. Recent studies have shown that adsorbed host serum components from normal mouse serum as well as Fc receptor associated normal mouse IgG can significantly reduce the amount of schistosome specific antibody binding at the parasite's surface (Rasmussen and Kemp, in press). This reduction is presumably due to stearic hindrance by the adsorbed components of antigen-antibody interaction and may be interpreted as supporting the concept of antigen masking of Smithers and Terry (1967). Furthermore, one is encouraged to speculate as to the possibility of a more subtle role which Fc receptor-bound anti- bodies might play. The orientation of these molecules is such that their role in camouflage of the parasite is partially negated by the prominent display of the only portion of the immunoglobulin which is foreign to the organism which made it, namely the idiotype. Manipulation of the idiotype network might give the parasite the edge necessary to keep the host slightly off balance immunologically in regard to itself without compromising the ability of the host to portect itself from other pathogens.

The possible functions of the C3 receptor are equally intrigu- ing. Stackpole et al. (1978) suggested that C3 receptors aided TL tumor cell immune escape by inserting themselves between cell surface antigen-antibody complexes. This would block complement activation by moving the antibodies from the necessary proximity for classical pathway activation to occur. Alternatively, Iida and Nussenzweig (1981, 1983) have shown that C3 receptors on red blood cells are capable of inhibiting C3 activation by blocking the enzymes necessary for C5 activation. These speculations become more reasonable when viewed in light of the work of Tsang et al. (1977) and Tsang and Damian (1977) which demonstrated that a parasite derived substance, termed bilharzin, is capable of binding to and inactivationg Factor XIIa of the clotting cascade (activated Hageman factor). Factor XIIa is the first step of the clotting process and therefore its inhibition blocks any chance of the process being activated. Likewise, although it is not the first component of the classical pathway, C3 is the first step in C5-9 proliferation and plays a significant role in initiating alterna- tive pathway activation.

One anomaly recently observed in our laboratory is that adult male worms incubated in normal mouse serum are damaged while adult male worms incubated in immune mouse serum are not damaged. Heat inactivation of the normal serum destroys the factors responsible for the damage, which suggests that this situation is similiar to alternative pathway activation by cercariae. Damage on the adult worm is limited to the tops of the dorsal tubercles (Rasmussen and Kemp, In Press). Current speculation is that the cercarial and adult stages express surface antigens capable of activating complement via the alternative pathway. Whether these chemicals are the same in both stages, and whether they are expressed only on these stages and either absent or hidden in some manner on schistosomula is unknown. It is tempting to speculate that these substances are suppressed during the migratory phases and are reexpressed and functional only when the parasites have developed enough to deal effectively with alternative pathway activation.

VIII.   SUMMARY

Our view of the mechanisms by which schistosomes protect themselves from host induced damage has shifted in recent years from a simplistic, one mechanism explanation to a concept where several complex and sophisticated mechanisms interplay and contribute to the parasite's survival.

Successful escape of a primary infection appears to be basically a once only phenomenon, taking advantage of a naive host, since subsequent invasions of larvae possessing the same escape mechanisms are usually failures. Adult worms on the other hand are well adapted to the host's immune response as evidenced by their longevity in the host and their ability to survive transplantation into an immunized host.

Speculation about adult mechanisms of escape at this time is that mimicry plays a role, either directly or indirectly, in reducing foreignness and alleviates some selection pressure of the host's response. Adsorbed host antigens inhibit the binding of the host's effector mechanisms, thereby alleviating immune-mediated damage. Classical complement pathway activation is thwarted by an inability of antibodies to bind to the adult worm, and C3 receptors are present to inactivate C3 if alternative pathway activation occurs or if antibodies manage to bind for a brief time. Finally, any breakdown in mimicry, host antigen adsorption, and/or complement inactivation mechanisms is temporary since the modulation process is present to rid the parasite of compromised or unwanted surface molecules before irreparable damage is done.

I hope that this brief treatment of the field of schistosome immune escape mechanisms has stimulated your interest and generated some enthusiasm for the area of parasite immunology. I suppose if I were to leave you with one "take home" message it would be best expressed by something that happened to me recently in the laboratory. A student asked if I knew the difference between "True Love" and "Herpes II". When I replied that I did not, I was informed that "Herpes II is forever". My closing comment is that Herpes II has nothing on schistosomes.

IX. REFERENCES

Auriault, C., Joseph, M., Dessaint, J. P., and Capron, A. (1980). Inactivation of rat macrophages by peptides resulting from cleavage of IgG by schistosoma larvae proteases. *Immunol. Letters* 2, 135–139.

Auriault, C., Ouaissi, M. A., Torpier, G., Eisen, H., and Capron, A. (1981). Proteolytic cleavage of IgG bound to the Fc receptor of *Schistosoma mansoni* schistosomula. *Parasite Immunol.*, 3, 33–44.

Capron, A., Biguet, J., Vernes, A., and Afchain, D. (1968). Structure antigenique des helminthes. Aspects immunologiques des relations host-parasite. *Pathol. Biol.*, 16, 121–128.

Caulfield, J. P., Korman, G., Butterworth, A. E., Hogan, M., David, J. R. (1980). The adherence of human neutrophils and eosinophils to schistosomula: Evidence for membrane fusion between cells and parasites. *J. Cell Biol.*, 86, 46–63.

Clegg, J. A., Smithers, S. R., and Terry, R. J. (1971). Acquisition of human antigens by *Schistosoma mansoni* during cultivation *in vitro*. *Nature*, 232, 653–654.

Cornford, E. M., Huot, M. E., Diep, C. P., and Rowley, G. A. (1982). Protein, glycogen, and water content in schistosomes. *J. Parasitol.*, 68, 1010–1020.

Damian, R. T. (1964). Molecular mimicry: Antigen sharing by parasite and host and its consequences. *Am. Nat.*, 98, 129–149.

Damian, R. T. (1967). Common antigens between adult *Schistosoma mansoni* and the laboratory mouse. *J. Parasitol.*, 53, 60–64.

Damian, R. T., Greene, N. D., and Hubbard, W. G. (1973).
    Occurrance of mouse alpha 2-macroglobulin antigenic deter-
    minants on *Schistosoma mansoni* adults, with evidence on their
    nature. *J. Parasitol.*, 59, 64-73.

Damian, R. T., Greene, N. D., Meyer, K. F., Cheever, A. W., Hubbard,
    W. J., Hawes, M. E., and Clark, J. D. (1976). *Schistosoma
    mansoni* in baboons III. The course and characteristics of
    infection, with additional observations on immunity. *Am. J.
    Trop. Med. Hyg.*, 25, 299-306.

Dean, D. A. (1974). *Schistosoma mansoni:* Adsorption of human
    blood group A & B antigens by schistosomula. *J. Parasitol.*,
    60, 260-263.

Dean, D. A. (1977). Decreased binding of cytotoxic antibody by
    developing *Schistosoma mansoni*. Evidence for a surface change
    independent of host antigen adsorption and membrane turnover.
    J. Parasitol., 63, 418-426.

Dean, D. A. and Sell, K. W. (1972). Surface antigens of
    Schistosoma mansoni. II. Adsorption of a Forssman-like antigen
    by schistosomula. *Clin. Exp. Immunol.*, 12, 525-540.

Dessein, A., Samuelson, J. C., Butterworth, A. E., Hogan, M.,
    Sherry, B. A., Vadas, M. A., and David, J. R. (1981). Immune
    evasion by *Schistosoma mansoni*. Loss of susceptibility to
    antibody or complement-dependent eosinophil attack by
    schistosomula cultured in medium free of macromolecules.
    Parasitology, 82, 357-374.

Gitter, B. D. and Damian, R. T. 91982). Murine alloantigen
    acquisition by schistosomula of *Schistosoma mansoni:* Further
    evidence for the presence of K, D, and I region gene products
    on the tegumental surface. *Parasite Immunol.*, 4, 383-393.

Gitter, B. D., McCormick, S. L., and Damian, R. T. (1982). Murine
    alloantigen acquisition by *Schistosoma mansoni:* Presence
    of H-2K determinants on adult worms and failure of allogeneic
    lymphocytes to recognize acquired MHC gene products on
    schistosomula. *J. Parasitol.*, 68, 513-518.

Hockley, D. J., and McLaren, D. J. (1973). *Schistosoma mansoni:*
    Changes in the outer membrane of the tegument during develop-
    ment from cercaria to adult worm. *Int. J. Parasitol.* 3,
    13-25.

Iida, K. and Nussenzweig, V.   (1981).   Complement receptor is an
    inhibitor of the complement cascade.   *J. Exp. Med.*, 153, 1138-
    1150.

Iida, K. and Nussenzweig, V.   (1983).   Functional properties of
    membrane-associated complement receptor CR1.   *J. Immunol.*,
    30, 1876-1880.

Kabil, S. M.   (1976).   Host complement in the schistosomal tegument.
    J. Trop. Med. Hyg., 79, 205-206.

Kemp, W. M.   (1970).   Ultrastructure of the Cercarienhüllen
    Reaktion of *Schistosoma mansoni*.   *J. Parasitol.*, 56, 713-723.

Kemp, W. M.   (1972).   Serology of the Cercarienhüllen Reaktion of
    Schistosoma mansoni.   *J. Parasitol.*, 58, 686-692.

Kemp, W. M., Greene, N. D., and Damian, R. T.   (1974).   Sharing of
    Cercarienhüllen Reaktion antigens between *Schistosoma mansoni*
    cercariae and adults and uninfected *Biomphalaria pfeifferi*.
    Am. J. Trop. Med. Hyg., 23, 197-202.

Kemp, W. M., Damian, R. T., Greene, N. D., and Lushbaugh, W. B.
    (1976a).   Immunocytochemical localization of mouse alpha 2-
    macroglobulin-like determinants on *Schistosoma mansoni* adults.
    J. Parasitol., 62, 413-419.

Kemp, W. M. Damian, R. T., and Greene, N. D.   (1976b).   Immuno-
    cytochemical localization of IgG on adult *Schistosoma mansoni*
    tegumental surfaces.   *J. Parasitol.*, 62, 830-832.

Kemp, W. M., Merritt, S. C., Bogucki, M. S., Rosier, J. G., and
    Seed, J. R.   (1977).   Evidence for adsorption of hetero-
    specific host immunoglobulin on the tegument of *Schistosoma
    mansoni*.   *J. Immunol.*, 119, 1849-1854.

Kemp, W. M., Merritt, S. C., and Rosier, J. G.   (1978).   *Schistosoma
    mansoni:* Identification of immunoglobulins associated with the
    tegument of adult parasites from mice.   *Exp. Parasitol.*, 45,
    81-87.

Kemp, W. M., Brown, P. R., Merritt, S. C., and Miller, R. E.
    (1980).   Tegument associated antigen modulation by adult
    male *Schistosoma mansoni*.   *J. Immunol.*, 124, 806-811.

McGuinness, T. B. and Kemp, W. M.   (1981).   *Schistosoma mansoni:*
    Evidence for a complement dependent receptor on adult male
    parasites.   *Exp. Parasitol.*, 51, 236-242.

McLaren, D. J., Clegg, J. A., and Smithers, S. R. (1975). Acquisition of host antigens by young *Schistosoma mansoni* in mice: correlation with failure to bind antibody *in vitro*. *Parasitology*, 70, 67-75.

McLaren, D. J., Hockley, D. J., Goldring, O. L., and Hammond, B. J. (1978). A freeze fracture study of the developing tegumental outer membrane of *Schistosoma mansoni*. *Parasitology*, 76, 327-348.

McLaren, D. J. and Incani, R. N. (1982). *Schistosoma mansoni:* Acquired resistance of developing schistosomula to immune attack *in vitro*. *Exp. Parasitol.*, 53, 285-298.

Moser, G., Wassom, D., and Sher, A. (1980). Studies of the antibody-dependent killing of schistosomula of *Schistosoma mansoni* employing haptenic target antigens: I. Evidence that the loss in susceptibility to immune damage undergone by developing schistosomula involves a change unrelated to the masking of parasite antigens by host molecules. *J. Exp. Med.*, 152, 41-53.

Quaissi, M. A., Santoro, F., and Capron, A. (1980). Interaction between *Schistosoma mansoni* and the complement system: Receptors for C3b on cercariae and schistosomula. *Immunol. Letters*, 1, 197-210.

Ritchie, L. S., Knight, W. B., Oliver-Gonzalez, J., Frick, L. P., Morris, J. M., and Croker, W. L. (1967). *Schistosoma mansoni* infections in *Cercopithecus sabacus* monkeys. *J. Parasitol.*, 53, 1217-1224.

Samuelson, J. C., Caulfield, J. P., and David, J. R. (1980). *Schistosoma mansoni:* Post-transitional surface changes in schistosomula grown *in vitro* and in mice. *Exp. Parasitol.*, 50, 369-383.

Santoro, F., Ouaissi, M. A., and Capron, A. (1979). Receptors for complement (C1z and C3b) on the immature forms of *Schistosoma mansoni*. *ICRS Med. Sci.*, 7, 576.

Santoro, F., Ouaissi, M. A., Pestel, J., and Capron, A. (1980). Interaction between *Schistosoma mansoni* and the complement system. Binding of C1q to schistosomula. *J. Immunol.*, 124, 2886-2891.

Sher, A., Hall, B. F., and Vadas, M. A. (1978). Acquisition of murine major histocompatibility complex gene products by schistosomula of *Schistosoma mansoni*. *J. Exp. Med.*, 148, 46-57.

Smithers, S. R. and Terry, R. J. (1967). Resistance to experimental infection with *Schistosoma mansoni* in rhesus monkeys induced by the transfer of adult worms. *Trans. Roy. Soc. Trop. Med. Hyg.*, 61, 517-533.

Smithers, S. R., Terry, R. J., and Hockley, D. J. (1969). Host antigens in schistosomiasis. *Proc. Roy. Soc. Lond. (Biol.)*, 171, 483-494.

Sogandares-Bernal, F. (1976). Immunoglobulins attached to and in the tegument of adult *Schistosoma mansoni* Sambon 1907, from first infection of CF1 mice. *J. Parasitol.*, 62, 222-226.

Stackpole, C. W., Jacobson, J. B., and Galuska, G. (1978). Antigenic modulation *in vitro*. II. Modulation of thymus leukemia (TL) antigenicity requires complement component C3. *J. Immunol.*, 120, 188-197.

Stirewalt, M. A. (1963). Cercaria vs schistosomula (*S. mansoni*): Absence of the pericercarial envelope *in vivo* and the early physiological and histological metamorphosis of the parasite. *Exp. Parasitol.*, 13, 395-406.

Tarleton, R. L. and Kemp, W. M. (1981). Demonstration of IgG-Fc and C3 receptors on adult *Schistosoma mansoni*. *J. Immunol.*, 126, 379-384.

Tavares, C. A. P., Gazzinelli, G., Mota-Santos, T. A., and Dias Da Silva, W. (1978). *Schistosoma mansoni*: Complement mediated cytotoxic activity *in vitro* and effect of decomplementation on acquired immunity in mice. *Exp. Parasitol.*, 46, 145-151.

Tavares, C. A. P., Cordeiro, M. N., Mota-Santos, T. A., and Gazzinelli, G. (1980). Artificially transformed *Schistosoma mansoni*: Mechanism of acquisition of protection against antibody-mediated killing. *Parasitology*, 80, 95-104.

Torpier, G., Capron, A., and Ouaissi, M. A. (1979). Receptor of IgF (Fc) and beta 2-microglobulin on *S. mansoni* schistosomula. *Nature*, 278, 447-449.

Tsang, V. C. W. and Damian, R. T.   (1977).   Demonstration and mode
    of action of an inhibitor for activated Hageman factor
    (Factor XIIa) of the intrinsic blood coagulation pathway from
    Schistosoma mansoni.   *Blood*, 49, 619-633.

Tsang, V. C. W., Hubbard, W. J., and Damian, R. T.   (1977).   Co-
    aggulation factor XIIa (activated Hageman factor) inhibitor
    from adult *Schistosoma mansoni*.   *Am. J. Trop. Med. Hyg.* 26,
    243-247.

Webbe, G., James, C., Nelson, G. S., Smithers, S. R., and Terry,
    R. J.   (1976).   Acquired resistance to *Schistosoma
    haematobium* in the baboon *(Papio anubis)* after cercarial
    exposure and adult worm transplantation.   *Ann. Trop. Med.
    Parasitol.*, 70, 411-424.

Wilson, R. A. and Barnes, P. E.   (1974).   An *in vitro* investigation
    of dynamic processes occurring in the schistosome tegument,
    using compounds known to disrupt secretory processes.
    Parasitology, 68, 239-258.

Wilson, R. A. and Barnes, P. E.   (1977).   The formation and turn-
    over of the membranocalyx on the tegument of *Schistosoma
    mansoni*.   *Parasitology*, 74, 61-74.

GENETIC CONTROL OF SCHISTOSOMIASIS: A TECHNIQUE BASED ON THE

GENETIC MANIPULATION OF INTERMEDIATE HOST SNAIL POPULATIONS

David S. Woodruff

Department of Biology
University of California, San Diego
La Jolla, California 92093

I.   INTRODUCTION

In this contribution I support the contention (Woodruff, 1978)
that it may be possible to reduce the size of human-infecting
schistosome populations by the genetic manipulation of their
intermediate host snails. The proposed technique is based on the
finding that snail-schistosome compatibility is variable and that
some of this variation is under relatively simple genetic regulation. If the proportion of intermediate host snails that are
resistant to infection by the local larval schistosome can be
increased, then the rate at which the parasite is transmitted to
the final host will decrease. Such genetically resistant snails
can be isolated by artificial selection procedures, mass-reared,
and their descendants returned to the population from which they
were isolated. If sufficient numbers of resistant snails are

41

released, the resultant genetic perturbation will be too great for
the local schistosome population to adjust coevolutionarily.
Under certain local ecological circumstances it may be possible
to break the transmission cycle within a few years.

At the outset, it must be stressed that this relatively simple
scenario for genetic control is not new; it first appears (in
simpler form) in the literature 30 years ago (World Health
Organization, 1954; Hubendick, 1958). Its potential has been
generally ignored, however, because of objections raised by some
early discussants. Here I will argue that these objections were
premature and that "the failure to adequately recognize intraspecific variability in response to infection has hindered experimental investigation of host-parasite relations and may have delayed the development of adequate control measures" (Wakelin,
1978).

II. GENETICS OF SCHISTOSOME-SNAIL COMPATIBILITY

Intermediate host specificity in the genus *Schistosoma* is
pronounced. Recent discussions of the 17 known species and their
intermediate host snails (Davis, 1980; Brown, 1980; Loker, 1983)
underscores the fact that, despite 100 million years of coevolution, *Schistosoma* spp. are able to avoid the defense mechanisms of
only about thirty species of gastropods. The evolution of the
various species groups of *Schistosoma* has been intimately linked
with the evolution of particular genera of snails. The schistosomes whose eggs bear a terminal spine (including *S. haematobium*
and *S. intercalatum*) are transmitted by about half of the twenty-
five known species of *Bulinus* (Pulmonata:Bulininae). Those
schistosomes whose eggs have a lateral spine (including *S.
mansoni*) are capable of developing in only about half of the
twenty-six species of *Biomphalaria* (Pulmonata:Planorbidae). The
Asian schistosomes, whose eggs lack prominent spines, are transmitted by only three of the 150 species of the prosobranch family
Pomatiopsidae (*S. japonicum* by *Oncomelania hupensis* and *S.
mekongi* by *Tricula aperta*). Almost without exception in any given
area a particular species of *Schistosoma* is capable of developing
in only a single species of snail. *S. mansoni*, for example, is
transmitted by *Biomphalaria alexandrina* in Egypt, by *B.
pfeifferi, B. sudanica* and several other species in different parts
of the sub-Saharan Africa, and by *B. glabrata, B. tenagophila,
and B. straminea* in different parts of Brazil (Wright, 1974; Brown,
1980). This pronounced, local intermediate host specificity constitutes a key factor in the proposed genetic control technique.

Intermediate host specificity is based on an overall compatibility between a local population of schistosomes and a
particular species of snail. Compatibility depends on both

parasite infectivity and snail susceptibility, and both genetic
and environmental factors effect the outcome. While the relative
significance of these various factors will vary within limits for
each miracidium-snail interaction, the limits themselves are set by
the underlying genetic determinants. These genes regulate the host-
parasite interaction through mechanisms that are still largely unknown; in compatible interactions the sporocyst develops without a
host tissue reaction, while in incompatible interactions the larval
schistosome is rapidly surrounded by amoebocytes and eventually
destroyed (Cheng, 1970; Basch, 1976; Harris, 1975; Bayne, 1982).

Present understanding of the genetics of snail susceptibility
is due primarily to the work of Richards (1970, 1973a,b, 1975a,b,c,
1976, 1977, 1983; Richards and Merritt, 1972; Lie et al., 1979).
Extending the work of Stunkard (1946) and Newton (1953, 1955),
Richards has established that the susceptibility of *B. glabrata* to
S. mansoni is under oligogenic control. [Oligogenic resistance,
also known as major gene resistance, is under the control of one
or a few genes whose individual effects are readily detected (Day,
1974)]. Challenging various laboratory strains of *B. glabrata*
with miracidia from strains of *S. mansoni* isolated in Puerto Rico
and St. Lucia, Richards has identified strains of contrasting
susceptibility, including: juvenile and adult resistant (Type I),
juvenile susceptible-adult resistant (Type II), juvenile and adult
susceptible (Type III). The latter, of course, is generally re-
garded as being the typical susceptibility type. Breeding and
selection experiments showed that in one strain challenged with
Puerto Rican miracidia, juvenile susceptibility is controlled by
at least four genes (Richards and Merritt, 1972) and adult susceptibility is determined by a single gene, with resistance being
dominant (Richards, 1973b). When the same snail strain was
challenged with miracidia from St. Lucia, juvenile susceptibility
was found to be determined by a single gene, with resistance being
dominant (Richards, 1975a). These studies indicate that, between
coevolved strains of host and parasite, compatibility is under
relatively simple genetic control, and that the exact nature of
this control varies with each host-parasite combination considered.
As in other cases of oligogenic resistance, the level of resistance
and the time at which it is expressed may vary greatly (Day, 1974).

Unfortunately, we still know comparatively little about the
genetics of schistosome infectivity. The tendency in the past was
to regard the schistosomes as being genetically homogeneous and to
assume that observed variations in compatibility were due entirely
to the snails. Wright (1974) correctly argued for a more balanced
view and pursued the problem of detecting variation in infectivity.
Using interspecific hybrid miracidia produced in the laboratory,
he found evidence for the inheritance of a snail infectivity factor.
He found that hybrids between *S. intercalatum* and *S. mattheei*,

which have mutually exclusive intermediate host snails, could
develop in both parental host snail species for at least three
generations. Similar observations of a hereditable infectivity
factor have been made by other workers who studied interspecific
or intraspecific hybrids in the laboratory (Files, 1951; Richards,
1975b; Wright and Southgate, 1976, 1981). These experiments also
show that hybrids are not equally compatible with both the parental
intermediate host snail stocks. In some cases (Wright, 1974;
Richards, 1975b) the hybrids were more compatible with the paternal,
in others (Files, 1951) with the maternal, schistosome host snail
stock. Useful as these studies are it will be logistically difficult to establish the formal genetics of these infectivity factors
using the hybrid schistosome technique.

Indirect evidence for the relatively simple genetic control of
susceptibility and infectivity comes from the ease with which one
can modify overall compatibility in the laboratory. Some examples
illustrate the point that both infectivity and susceptibility may
be altered by deliberate or inadvertant selection. Saoud (1965)
found that an Egyptian strain of *S. mansoni* that had been passaged
through American *B. glabrata* for 15 years was almost noninfective
to its natural host, *B. alexandrina*. McClelland (1965) describes
the inadvertant selection of a strain of *Bulinus nasutus* resistant
to *S. haematobium* over a 4-year period. Kagan and Geiger (1965)
inadvertantly raised levels of compatibility between several strains
of *S. mansoni* and *B. glabrata* over a 3-year period by systematically
discarding resistant snails. Richards and Merritt (1972) report a
Puerto Rican stock of *B. glabrata* that became resistant to *S.
mansoni* while in the laboratory. Richards (1975b) was able to
select two lines of *S. mansoni* from his St. Lucia strain that
differed in their infectivity.

More recently, Santana et al. (1978) have shown how deliberate
selection procedures involving collecting the self-fertilized offspring of susceptible snails and discarding resistant snails produces very rapid shifts in the overall susceptibility of a
population. In one experiment they increased the susceptibility of
Biomphalaria tenagophila from near Sao Paulo, Brazil, to sympatric
S. mansoni from a typically low value of 7% to 97% in two generations. In a second experiment involving *B. glabrata* and *S. mansoni*
from Belo Horizonte, susceptibility increased from 40% to 93% in a
single generation. In both cases the increased susceptibility was
maintained in subsequent generations studied. While formal genetic
interpretations of these observations are impossible a posteriori,
the speed with which the changes occurred indicates that relatively
few genes are involved.

Such rapid shifts in compatibility even have been observed in nature, albeit in an ecosystem perturbed by man. Wright and Southgate (1979) discuss a study of Duke and Moore (1976a,b) in which *S. haematobium* in a small lake in Cameroon switched intermediate host snails over a 4-year period. The situation was most unusual in that *S. haematobium* was originally transmitted by both *Bulinus rohlfsi*, its normal host, and to a minor extent by *Bulinus camerunensis*, a host of *S. intercalatum*. This dual infectivity is probably a legacy of a previous period of hybridization between *S. haematobium* and *S. intercalatum* in this area (Wright et al., 1974). The host shift occurred when repeated applications of a molluscicide proved far more effective against *B. rohlfsi* than against *B. camerunensis*. While the treatments virtually eliminated transmission through *B. rohlfsi*, transmission through *B. camerunensis* increased to three times the precontrol level. Clearly, the selection pressures favored parasites carrying *S. intercalatum* infectivity genes.

There is also some circumstantial evidence for the gradual change in levels of compatibility in nature in areas where the range of the parasite has increased as a result of human migration. In southeast Brazil, for example, regional differences in intermediate host-parasite compatibility may be interpreted as reflecting varying degrees of coevolution (Bastos et al., 1978a; Paraense and Correa, 1978; Cavalho et al., 1979; Correa et al., 1979). Studies of mortality rates and numbers of cercariae released from infected snails indicate that *S. mansoni* is better adapted to *B. glabrata* than to *B. tenagophila*, whose range it has more recently invaded. Furthermore, in some areas where schistosomiasis has only recently become a health problem. *B. tenagophila* is apparently more compatible with the strain of the parasite cycling through local wild rodents than it is to the strain of parasite found in local people (Bastos et al., 1978a,b). Earlier studies showing that a strain of *S. mansoni* adapted to *B. tenagophila* would not infect *B. glabrata* and vice versa (Paraense and Correa, 1963) have not been confirmed by more recent work (Bastos et al. 1978a; Carvalho et al., 1979). Whether compatibility levels have changed during the intervening period or whether the differences are due to experimental procedures should be established. A similar situation may prevail in Angola (Morais, 1978) and the Philippines (Rachford, 1977).

A critical study of the genetic variation in compatibility within a single coevolved population of schistosomes and snails has yet to be reported (Frandsen, 1979). This is surprising because it has long been known that the prevalence of infection in snail populations is typically very low, and the obvious possibility that some of the uninfected snails must be resistant must have occurred to many people. Empirical (Etges, 1963; Chu et al., 1966)

and theoretical (Donges, 1974) arguments that snails are homogeneous
with respect to susceptibility have been refuted by Basch (1975)
and Clarke (1979) who have shown that heterogeneity is to be expected in both snails and parasites. Nevertheless, data on the distribution and frequency of the genes affecting compatibility at a
transmission site are simply nonexistent. Field prevalence data
are of little use as the environmental components of their variation
cannot be eliminated. Similarly, we cannot estimate gene frequencies
from infection rates obtained in the laboratory, because, typically,
each snail may be exposed to only a fraction of the schistosome gene
pool. We might utilize the ecologically important fact of parasite
overdispersion in the number of miracidial infections per snail
to measure differences in compatibility as suggested by Anderson
(1978a). The relationship between the variance to mean ratio of
the number of infections per snail and miracidial density provides
a framework for the quantification of within-strain variability,
even though it tells us nothing about its genetic control.

    In summary, the limited evidence available indicates that both
susceptibility and infectivity are under oligogenic control and that
different interactive gene combinations have evolved in each local
host-parasite association. The rapid responses to selection for
different levels of compatibility, couped with Richard's analyses
of the formal genetics of susceptibility, suggest that dominance
and additive effects are important in determining the genetic variance in each host-parasite combination. The probable lack of a
single species-wide genetic system for each host-parasite combination is suggested by the poor correlation between experimentally
obtained infection rates and the geographic distance separating the
populations from which the snail and schistosome were derived (Lo,
1972; Davis and Ruff, 1973; Basch, 1976). These conclusions are
not unexpected; analogous simple genetic mechanisms have been
found to control the susceptibility of mosquitoes to filarial worms
and malarial parasites, and the susceptibility of leafhoppers to
arboviruses (Davidson, 1974; MacDonald, 1976; Wakelin, 1978).
It may be that much invertebrate "immunity" is regulated by relatively simple oligogenic mechanisms.

III. THE GENETIC CONTROL TECHNIQUE

    If the foregoing conclusions on the genetic regulation of com-
patibility are correct, then it should be possible to select snails
that are resistant to some or all the infectivity types of the
schistosome population they normally transmit. The genetic control
technique would involve mass-rearing such resistant snails and
returning their descendants in large numbers to the transmission
site from which they were derived. The resulting genetic perturbation decreases the rate at which the larval schistosomes pass

through the snail population, and in due course decreases the
frequency of new infections in the definitive host population.

There are at least three ways of developing a resistant
strain of snails for use in the genetic control of schistosomes.
First, one could select for incompatibility within a local population of snails at a disease transmission site. Second, one could
import host snails from another area if it were known that the
imported snails were incompatible with the local strain of the
parasite. Third, one could hybridize local compatible snails with
conspecific but incompatible snails from elsewhere and use those
hybrids resistant to the local schistosomes.

The first method has the advantage of involving snails that
should be better adapted to conditions at the release site than
imported snails or their hybrids. This point has been made previously by Thomas (1973) who, after considering the development of
a resistant strain by the second method only, rejected the possibility of achieving genetic control.

The other two methods have an obvious procedural advantage in
that the time-consuming initial selection for resistance genes is
avoided. Nevertheless, until more is known about the genetic regulation of compatibility, I advocate using the first method. I
agree with Wright (1974) that the genetic basis of compatibility
probably varies from place to place. This is precisely what we
would expect on the basis of better known cases of a species's
genetic response to a parasite; in the wheat rust wheat interaction
resistance is under the regulation of a fairly simple genetic
system, the precise details of which vary geographically (Day, 1974).
If compatibility has evolved to a greater or lesser extent in each
local population then it is most meaningful to study it in the
context in which it evolved-the local population.

Until genetic or other markers linked with compatibility
states are identified, the selection of resistant snails will be a
tedious process. Field-collected snails must be isolated singly
or in pairs and screened repeatedly for cercarial output over a
period of at least 8 weeks. Eggs (self-fertilized in the case of
the pulmonates) of apparently unparasitized snails are then isolated to avoid echinostome transmission and metacercarial infection,
and are used to establish the first laboratory generation. Cohorts
of first generation snails are then tested individually for
resistance to miracidia derived from parasitized members of the
parental snail population. It is important that snails be exposed
in a standard manner to as much of the spectrum of local schistosome
infectivity as possible, i.e., there is little merit in selecting
for resistance against only a small fraction of the parasite genome.
Snails must be exposed as juveniles and then, in the absence of

infection, again as adults. The offspring (self-fertilized when possible) of resistant adult snails then are used to found the second laboratory generation. The process is repeated in subsequent generations until a true-breeding resistant stock, or one with suitably low susceptibility, is derived. This is a tedious and logistically complex process, but work in my laboratory suggests that it is not impossible.

Once a suitably incompatible strain of snails has been established there should be little difficulty in obtaining the large numbers of snails required for liberation at the transmission site. Ecologically, most species of snails that serve as intermediate hosts of schistosomes are r-strategists and have a very high propensity for rapid population growth. Under favorable laboratory conditions, a pair of snails can give rise to several thousand offspring in a year. Mass-culturing techniques have been developed for most of the medically important species (Bruce and Radke, 1971; Davis, 1971; Liang, 1974).

If adequate numbers of resistant snails are liberated at the site of disease transmission and successfully reestablish themselves in nature, they will have the following effects:

1. There will be a sudden decrease in the proportion of susceptible snails and a concomitant increase in the proportion of resistant snails. These changes are simply a function of the release ratio: the relative numbers of genes for resistance in the released population and in the original population.

2. As miracidia are unable to distinguish resistant from susceptible snails, the former act as decoys and trap a portion of the schistosome population.

3. Parasitized susceptible snails are less fit in an evolutionary sense than are resistant snails; they show greatly reduced fecundity and longevity. On-going natural selection will therefore favor the genes for resistance as long as parasites are present.

4. Hybridization between introduced and endemic snails will result in a further increase in the frequency of the genes for resistance.

These effects are predicted on the basis of empirical evidence and evolutionary theory. While the origin of the initial change in gene frequencies is of course obvious and needs no further explanation, the other effects deserve comment. The "decoy effect" is based on considerable evidence that miracidia do not discriminate between compatible and incompatible snails (Stunkard, 1946;

Newton, 1952; Barbosa and Barreto, 1960; Sudds, 1960; Upatham, 1972; Upatham and Sturrock, 1973). In fact, there are cases where miracidia were found to penetrate a resistant species of gastropods more frequently than a closely related susceptible one (Lo Verde, 1976). The use of this effect alone has been the basis of one schistosome control proposal (Laracuente et al., 1979).

The third effect, and perhaps the most important, is based on the observation that "schistosome fever" is generally fatal to the snail. While infected snails may cease to release cercariae if kept in suboptimal conditions, the evidence that they can ever fully recover is not compelling (Anderson et al., 1977; Anderson and May, 1979). The pathology associated with the parasite's growth, migration through the tissues, and asexual reproduction shorten the snail's longevity. Death rates of infected snails are typically 1.3-4.0 times higher than those of uninfected snails. Infection tends to reduce the life expectancy of the snail by a factor of three (Hairston, 1973; Anderson and May, 1979). Infection also brings about a cessation of reproduction. Fletcher (1984) analyzed data provided by Sturrock (1966) and found the net reproductive rate (R_o) in *B. pfeifferi* was 1693.6 in uninfected snails and only 26.8 in infected snails. Those snails that are uninfected because they possess genes for resistance have a considerable selective advantage over infected snails.

The fourth and final effect is based on the supposition that mating between introduced and endemic snails will occur at random. Although it is not yet possible to investigate this effect experimentally, several lines of evidence support this prediction. Richards's breeding experiments indicate that snails of different susceptibility types and their hybrids are fully compatible with one another. The fact that miracidia are unable to distinguish between compatible and resistant snails suggests that the snails themselves also may be unaware of this variation and will accordingly mate at random. Finally, behavioral studies indicate that outbreeding is the norm, even in those species that are facultative hermaphrodites (Newton, 1953; Paraense, 1955). This latter conclusion is confirmed by electrophoretic studies of allozyme genotype frequencies in field-collected *B. alexandrina*, *B. obstructa*, *B. straminea*, and *B. glabrata* (Mulvey and Vrijenhoek, 1981, 1982; Woodruff et al., unpubl.). The net result of these effects will be a progressive increase in the frequency of the genes for resistance in the local snail population. The process would be analogous to that of the well-documented cases of transient polymorphisms involving the rise of the melanic form of the peppered moth, *Biston betularia*, in industrial England (Bishop and Cook, 1975) and the rapid evolution of pesticide resistance in various insects (Plapp, 1976).

The magnitude and rate of these various effects might be
estimated by means of mathematical models and simulations. Fletcher
(1984) provides an example of the form such an analysis might take.
While she emphasizes that quantitative simulations are premature,
her deterministic model of transmission of *S. mansoni* by *B. glabrata*
in a closed ecosystem, e.g., an oasis or island with no parasite
migration, illustrates the principles of the control process.
Based on findings (described above) that two cases of adult resistance were determined by single genetic factors with resistance
being dominant, Fletcher treated susceptibility as a monogenic
trait and assumed that the parasites were uniformly infective. By
treating the fitnesses of the different susceptibility phenotypes,
the initial infection rate (in the snails), the release ratio, and
the frequency of releases as variables, she was able to follow the
change in incidence in the snail population and the change in the
frequency of the genes for susceptibility. When repeated releases
are made at a relatively high ratio, e.g., 1:1, once a year, the
infection rate decreases substantially while the frequency of the
resistant gene increases until it reaches very high levels. Control
is most effective when initial snail infection rates are high and
when resistant snails have a selective advantage over susceptible
snails. Under the conditions of her model, she found that these
changes occur rapidly over the first 1 to 4 years of the simulation
in most of the circumstances considered. She found, to give just
one example, that if resistance is dominant it should be possible
to reduce the snail infection rate from 1 to 10 to 1 in 10,000 in
1 year by introducing resistant snails at a ratio of 1:1 four times
during the year. If the release ratio was only 1:10 the same change
takes about 5 years.

With increasing numbers of miracidia unable to develop successfully in the intermediate host snail population, the number of
cercariae at the transmission site will decrease. The incidence
of new infections in the final host will also decline and the adult
schistosome population will begin to decrease. In open ecosystems
the perturbation may at least reduce the incidence rate in the final
host for a few years. In a closed ecosystem, if the perturbation
is severe enough, the schistosome population may not be able to
respond fast enough to avoid eventual extinction. The principal
reason for this is not that schistosomes lack the genetic propensity
to counter the perturbation but that they may not be able to
evolve fast enough. Evolutionary rates, which are related to an
organism's generation time, are different in parasite and host.
While individual snails may live more than a year, mean longevity
of most species is 25-60 days and the minimum generation time at
25°C is approximately 42 days for *B. glabrata*, 75 days for *O.*
hupensis, and 63 days for *B. truncatus* (Otori et al., 1956; Pesigan
et al., 1958; Hairston, 1965; Davis and Iwamoto, 1969; Sturrock,
1973; El-Hassan, 1974). In contrast, individual worms may live up

to 33 years, mean longevity of adults is probably 3-5 years, and minimum generation time at 25°C is approximately 90 days for *S. mansoni*, 105 days for *S. japonicum*, and 130 days for *S. haematobium* MacDonald, 1973; Warren et al., 1974; Anderson and May, 1979; Market et al., 1979; Loker, 1983). These data suggest that the schistosomes have minimum generation times about twice those of their host snails. This fact, coupled with their much greater longevity as reproducing adults, will limit their ability to respond quickly to the genetic perturbation in the local snail population.

IV. DISCUSSION

In focusing on the intramolluscan stage of the parasite's life the genetic control technique seeks to reduce transmission at a point in the cycle of known sensitivity to control (MacDonald, 1965; Jordan, 1977; Webbe, 1978; Bradley, 1983). In this section I discuss some of the possible advantages of the genetic control technique over existing methods of snail control, enumerate some known problems that impede immediate application of the technique, and discuss problems raised by critics of the technique's potential. At the outset, I note that some of these criticisms have arisen simply because different parasitologists have given different meanings to the term genetic control. Some have used this term to denote the underlying genetic regulation of invertebrate incompatibility (Van den Bosch and Messenger, 1973; Barr, 1975; Walelin, 1978). Others have implied pest population control by the direct genetic manipulation of the target species as in the case of the sterile male technique (Davidson, 1974; Pal and Shitten, 1974). Here, I employ the term in a third sense for the technique of reducing the size of a pest species population by the genetic manipulation of another species, in this case its intermediate host. In this usage I follow Richards (1970), who has set the stage for the technique's development.

The genetic control technique is based on the pronounced inter-mediate host specificity that characterizes the blood flukes of the genus *Schistosoma*. By concentrating on the larval schistosome, one does not have to be concerned with the "reservoirs" of adult worms in mammals other than man. For example, medically important schistosomes occur in rats in Egypt (Mansour, 1973), Brazil (Coelho et al., 1979), and Guadeloupe (Theron et al., 1978), in primates in East Africa (Jordan and Webbe, 1969) and Ethiopia (Fuller et al., 1979), and in rats and dogs in the Philippines (Hairston, 1962). For a review, see Loker (1983). Some existing control projects still ignore the fact that human schistosomes may continue to cycle and pose a threat, even if the adult worms could be eliminated from the present human population.

This proposed technique of genetic control has a number of advantages over existing methods of snail control. Until recently, the principal method of snail control involved the use of synthetic and natural molluscicides (Malek, 1978, 1980). While undoubtedly very effective in some situations, the technique is inappropriate for many transmission sites and it is becoming too expensive for widespread application (Webbe, 1978). In addition, it has several major biological shortcomings. First, snails tend to avoid toxic compounds such as molluscicides; in some cases treated adults are known to climb out of the water and estivate (Uhazy et al., 1978). Second, snails will eventually evolve resistance to the molluscicides in widespread use, just as insects quickly developed resistance to insecticides. This phenomenon does not appear to have been appropriately monitored by malacologists; the reports by Walton et al. (1958), Yasuraoka (1972), Jelnes (1977) and Sullivan et al. (1984) appear to be alone in the field. Third, snails are important members of quatic ecosystems; it is ecologically unsound to attempt to destroy these organisms that are important in energy flow and nutrient cycling (Cummins and Klug, 1979). Fourth, molluscicides and their derivatives have largely unknown effects on non-target species including man and his food organisms. Finally, as originally conceived, mollusciciding is based on a faulty biological premise: that the host snail population should be eradicated. Not only is it virtually impossible to eradicate organisms with such good dispersal abilities, but it is also an inappropriate management goal when applied to r-strategists (Kuris, 1973). As Pesigan et al. (1958) found, it took only 8 months for a population of *O. hupensis* to recover from an 85% reduction in numbers. The populations of such opportunistic species are typically regulated in nature by the weather and environmental catastrophies, and their populations recover rapidly from each setback. More importantly, attempts to eradicate the intermediate host may result ultimately in both snail and parasite populations and may afford the parasite even further protection by virtue of its increased numbers (Anderson, 1978b).

Alternative nonchemical methods of snail control have consequently received considerable attention. Berg (1973) and Hairston et al. (1975) have reviewed the growing literature on biological control technique based on introducing predators, competitors, or larval antagonists. As entomologists can testify (De Bach, 1974), these techniques, though attractive, sometime produce highly unpredictable consequences. To date, however, these techniques of snail control have met with only limited success and they remain to be proven as major tools of control outside of Puerto Rico (Jobin et al., 1977; Hoffman et al., 1979; Jobin and Laracuente, 1979; Laracuente et al., 1979). In fact, it remains to be demonstrated that natural populations of schistosome host snails are regulated to any significant extent by their enemies. As the history of malaria control demonstrates (Harrison, 1978), a similar situation occurs with mosquitoes.

By reiterating the problems of existing control technique that focus on the intramolluscan stage of the schistosome's life, I do not mean to imply that the genetic control technique is without limitations. While it avoids many of the above mentioned problems by not seeking to eradicate or even to reduce the number of snails, it is not without problems of its own. In the following discussion I will consider some technical difficulties that have already been recognized, and will suggest ways in which they might be overcome. I will also discuss the reservations of the critics of the genetic control technique.

The reader with any field experience with schistosome transmission will undoubtedly have thought of some situation where the genetic control technique may not work. Each host-parasite combination and each transmission site has its own special characteristics; post-perturbation decreases in schistosome transmission rates will be slower in some circumstances than in others. Consideration of the known facts about compatibility, mating systems, and generation times of both flukes and snails suggests that the *S mansoni-Biomphalaria* spp. association provides the best system for the technique development and assessment. Relatively isolated transmission sites, where schistosome migration is low, would be more appropriate for early research than would larger, more open areas. We cannot really assess the technique's potential in any given area until we know something about the local distribution and frequency of the genes for compatibility in both snails and schistosomes.

At this stage the proposed technique's greatest weakness is its genetic basis; the precise extent of genetically controlled variation in compatibility still has to be established for a transmission site. While laboratory studies indicate that resistance and infectivity are under oligogenic control, this must now be shown for a natural host-parasite association. If this relatively simple genetic underpinning can be confirmed, the genetic control technique should work along the lines outlined above and by Fletcher (1984). Reservations expressed by Harrison et al. (1975) that the introduced population (which they envisioned as being derived from elsewhere and at some local selective disadvantage) will be swamped genetically by the local strain are no longer valid if the resistant snails are derived locally. Similarly, Michelson and Dubois's (1978) concern that the mixed population would develop only an intermediate level of susceptibility is not supported by our present understanding of the genetics of compatibility.

A second technical problem requiring attention involves the
artificial selection process. Selection will undoubtedly affect
more than just the target susceptibility loci and it is possible
that a prolonged laboratory confinement will reduce the snails'
fitness when they are returned to nature. This problem was raised
by Thomas (1973) and led him to speculate that the technique might
not succeed. This objection would presumably hold for a resistant
strain like Richards's Type I B. *glabrata* which has been in the
laboratory for over 10 years, but should be largely negated by the
selection procedure proposed above where snails are returned to
their ancestral habitat within 2 to 3 years. There is certainly no
need to confine more than the first few generations of selected
snails to the laboratory; mass-rearing could be conducted in a semi-
natural setting.

There is some prospect that the selection process could be
speeded up considerably. If biochemical or other correlates of
resistance can be found, it is conceivably possible to reduce the
selection process to a single laboratory generation. The search
for such a marker is hampered by our lack of a clear understanding
of the cellular and biochemical processes by which the genetic control of compatibility is mediated. An understanding of these
processes might allow us to score a snails compatibility directly,
on the basis of some characteristic like the presence or absence of
a particular allozyme or the level of concentration of a certain
substance in the hemolymph of snails challenged with parasite antigens. Clearly, the regulation of compatibility deserves more
attention.

The utility of a resistant strain of snails beyond the confines
of the site from which it was derived is conjectural. Hoffman
et al. (1979) state that compatibility is highly variable and express
the opinion that studies done in one place would be of little use in
day-to-day control efforts elsewhere. This reservation is premature,
as there is absolutely no information on the geographic distribution
of compatibility genes in either host or parasite. Cross-infectivity
studies, where snails from different areas are challenged with a
single parasite stock, and local adult snail infection rates, tell
us nothing about the genes affecting compatibility or their distribution. They do not even tell us much about relative levels of compatibility, because within-strain variation is typically ignored.
While it might be argued that the highly focal nature of schistosome
distribution favors the evolution of strictly localized host-parasite
compatibility, there are as yet no data on this point and further
speculation is unwarranted.

A widely held reservation about the genetic control technique
was raised by Wright (1971b) in his review of Richards (1970)
original proposal. If such an approach is likely to become effective,

how is it that populations of snails occur in nature which successfully transmit parasites and yet include nonsusceptible
individuals? Why has insusceptibility not become dominant in
such places? The full answer to this question cannot be made in
the absence of data on the frequencies of resistance genes in nature
and the relative fitness of the various susceptibility phenotypes
at a transmission site. At this stage, I can only follow Clarke
(1979) and speculate that the parasites simply do not exert enough
pressure on the snail population to bring the genes for resistance
to high frequency. Complete resistance has not evolved because
the parasites infect too small a proportion (typically 1% of the
adults) of the snails. Each parasite species is well adapted to
its host species despite the high pathogenicity seen in a few
individual snails; the two have coevolved a mutual accommodation.
This explanation does not necessitate that resistance entails some
biochemical price, reduced fecundity or physiological defect as
postulated by Wright (1971b), Hairston et al. (1975), and partici-
pants in a recent workshop (Fine, 1976). Only very recently has
there been any real data to support this postulate: Minchella
(1983) reports finding that unsuitable insusceptible snails were
negatively affected in the presence of either susceptible snails
or schistosomes. His two laboratory experiments involve highly
aberrant snail stocks (both susceptible and unsuitable stocks were
derived from a single parent) and ignore juveniles in the preparation of the partial life tables. Nevertheless, more studies of
this type are need before his finds on the cost of unsuitability
(not "resistance" as used in this paper) can be generalized. In
passing, I point out that even if the resistant snails were found
to be slightly less fit than susceptible ones, they can still be
employed effectively in the genetic control technique (Fletcher,
1984).

     A second major reservation involves the nature and consequences
of the schistosome's response to the genetic perturbation. In the
normal course of events the alteration of a single parameter in
the schistosome's indirect life cycle can presumably be compensated
for by the alteration of one of many other parameters (Anderson,
1974; Loker, 1983). For example, a change in intermediate host
susceptibility might be countered by changes in miracidial host
discrimination, infectivity, or pathogenicity, thus allowing the
association to recover before extinction occurs. I have argued
that the genetic control technique should involve perturbations
of such magnitude and suddenness that the parasite is unable to
respond. While it is difficult to imagine how the parasite's
compensation reactions can be triggered fast enough to avoid a
decrease in transmission, we must anticipate a strong reaction.
Thomas (1973) has argued that the perturbation might result in
previously refractory snails becoming susceptible. Wright
(1968, 1971a) has argued that the introduced snails may be more

susceptible to some other parasite and that it could spread and cause new problems elswhere. While there are no data to support these reservations and the risks should be minimized by the use of conspecific locally-derived snails, any strain of snails developed for control purposes must be examined for these risks before field liberation.

I conclude that the prospects for the genetic control technique have been ignored or down played for 25 years as a result of numerous ill-founded and premature speculations. This is a sad commentary on the state of medical malacology, a field which still lags far behind entomology in the application of the principles of population ecology and genetics to control problems (Anderson, 1982; Anderson and May, 1982; Woodruff, 1983). The main purpose of this review is therefore to draw attention to the scientific issues remaining to be tackled before the genetic control technique can be fairly assessed. Undoubtedly, the further study of the genetic regulation of host-parasite compatibility will be helpful in combating schistosomiasis and other snail-borne diseases, whether or not this particular control technique proves meritorious. As genetic engineering techniques will soon be available that would facilitate the development of resistant strains of appropriate fitness, it would seem highly desirable to improve our understanding of the genetic basis of compatibility at this time.

V. ACKNOWLEDGEMENTS

Charles S. Richards's (1970) paper on the genetics of *Biomphalaria glabrata* provided the inspiration for this study of the genetic control technique, which I undertook at Purdue University, Indiana, during the period 1976-79. The project has benefited immeasurably from his support. I thank my former students, Madeleine Fletcher, Michael Goldman, and Dennis Minchella, for their numberous contributions. In developing the ideas expressed in this paper, I have enjoyed stimulating discussions with many individuals including D. J. Bradley, J. B. Burch, B. C. Clarke, J. E. Cohen, G. M. Davis, W. M. Hirsch, R. C. Lewontin, R. M. May, and K. S. Warren. I am indebted to the Edna McConnell Clark Foundation for a grant that made this work possible; the importance of their funding a number of rather speculative research projects which, like the present study, involved nontraditional approaches to schistosomiasis control, cannot be overestimated.

VI. REFERENCES

Anderson, R. M. (1974). Population dynamics of the cestode
  Caryophyllaeus laticeps (Pallas, 1781) in the bream
  (*Abramis brama* L.). *J. Anim. Ecol.*, 43, 305-321.

Anderson, R. M. (1978a). Population dynamics of snail infection
  by miracidia. *Parasitology*, 77, 201-224

Anderson, R. M. (1978b). The regulation of host population growth
  by parasitic species. *Parasitology*, 76, 119-157.

Anderson, R. M. (ed.) (1982). "Population Dynamics of Infectious
  Diseases". Chapman and Hall, London.

Anderson, R. M. and May, R. M. (1979). Prevalence of schistosome
  infections within molluscan populations: observed patterns and
  theoretical predictions. *Parasitology*, 79, 63-94.

Anderson, R. M. and May, R. M. (eds.). (1982). "Population
  Biology of Infectious Diseases." Springer-Verlag, New York.

Anderson, R. M., Whitfield, P. J., and Mills, C. A. (1977). An
  experimental study of the population dynamics of an ecto-
  parasitic digenean *Transversotrema patialense* (Soparker):
  the cercarial and adult stages. *J. Anim. Ecol.*, 46, 555-580.

Barbosa, F. S. and Barreto, C. (1960). Differences in susceptibility of Brazilian strains of *Australorbis glabratus* to
  Schistosoma mansoni. *Exp. Parasitol.*, 9, 137-140.

Barr, A. R. 1975. Evidence for genetical control of invertebrate
  immunity and its field significance. *In* "Invertebrate
  Immunology" (K. Maramorosch, and R. E. Shope, eds.)
  Academic Press, New York.

Basch, P. F. 1975. An interpretation of snail-trematode infection rates: specificity based on concordance of compatible
  phenotypes. *Int. J. Parasitol.*, 5, 449-452.

Basch, P. F. 1976. Intermediate host specificity in *Schistosoma
  mansoni*. *Exp. Parasitol.*, 39, 150-169.

Bastos, O. de C., Guaraldo, A. M. A., and Magahlaes, L. A. (1978a).
  Suscetibilidade de *Biomphalaria glabrata*, variante albina,
  oriunda de Belo Horizonte, MG, a infeccao por *Schistosoma
  mansoni*, parasita em codicoes naturais, de roedores
  silvestres do Vale do Rio Paraiba do Sul, SP (Brasil). *Rev.
  Saude Publ.*, *Sao Paulo*, 12, 179-183.

Bastos, O. de C., Magahlaes, L. A., Rangel, H. de A., and Pidrabuena, A. E. (1978b). Alguns dados sobre o comportamento parasitologico das linhagens humana e silvestre do *Schistosoma mansoni*, no Vale do Rio Paraibaa do Sul, SP (Brasil). *Rev. Saude Publ., Sao Paulo*, 12, 184-199.

Bayne, C. J. (1982). Recognition and killing of metazoan parasites, particularly in molluscan hosts. *In* "Developmental Immunology: Clinical Problems and Aging." pp. 109-114, Academic Press, New York.

Berg, C. O. (1973). Biological control of snail-borne diseases: a review. *Exp. Parasitol.*, 33, 318-330.

Bishop, J. A. and Cook, L. M. (1975). Moths, melanis and clean air. *Sci. Am.*, 232, 90-99.

Bradley, D. J. (1982). Epidemiological models - theory and reality. *In* Anderson (1982) pp. 320-333.

Brown, D. S. (1980). "Freshwater Snails of Africa and Their Medical Importance." Taylor and Francis, London.

Bruce, J. I. and Radke, M. G. (1971). Cultivation of *Biomphalaria glabrata* and maintenance of *Schistosoma mansoni* in the laboratory. *Bio-Medical Rep. 406th Med. Lab.*, 19, 1-84.

Carvalho, O. S., Milward-De-Andrade, R., and Souza, C. P. (1979). Susceptibilidade de *Biomphalaria tenagophila* (d'Orbigny, 1835), de Itajuba (MG), a infeccao pela cepa "LE" de *Schistosoma mansoni* Sambon, 1907, de Belo Horizonte, MG (Brasil). *Rev. Saude Publ., Sao Paulo*, 13, 20-25.

Cheng, T. C. (1970). Immunity in mollusca, with special reference to reactions to transplants. *Transplant. Proc.*, 2, 226-230.

Chu, K. Y., Sabbaghian, H., and Massoud, J. (1966). Host-parasite relationship of *Bulinus truncatus* and *Schistosoma haematobium* in Iran. 2. Effect of exposure dosage of miracidia on the biology of the snail host and the development of the parasites. *Bull. WHO*, 34, 121-130.

Clarke, B. C. (1979). The evolution of genetic diversity. *Proc. R. Soc. London* (B), 205, 453-474.

Coelho, P. M. Z., Diaz, M., Mayrink, W., Magahlaes, P., Mello, M. N., and Costa, C. A. (1979). Wild reservoirs of *Schistosoma mansoni* from Caratinga, an endemic schistosomiasis area of Minas Gerias State, Brazil. *Am. J. Trop. Med. Hyg.*, 28, 163-164.

Correa, M. C. dos R., Coelho, P. M. Z., and Freitas, J. R. (1979). Susceptibilidade de linhagen de *Biomphalaria tenagophila* e *B. glabrata* a duas cepas de *Schistosoma mansoni* (LE-Belo Horizonte: M.G., SJ-San Jose dos Campos, SP). *Rev. Inst. Med. Trop. Sao Paulo*, 21, 72-76.

Cummins, K. W. and Klug, M. J. (1979). Feeding ecology of stream invertebrates. *Ann. Rev. Ecol. Syst.*, 10, 147-172.

Davidson, G. 1974. "Genetic Control of Insect Pests." Academic Press, New York.

Davis, G. M. (1971). Mass cultivation of *Oncomelania* (Prosobranchia: Hydrobiidae) for studies of *Schistosoma japonicum*. *Bio-Medical Rep. 406th Med. Lab.*, 19, 85-161.

Davis, G. M. (1980). Snail hosts of Asian *Schistosoma* infecting man: evolution and coevolution. *In* "The Mekong Schistosome" (C. Harinasuta, and J. L. Bruce, eds.) Univ. of Michigan Publ. Ann Arbor.

Davis, G. M. and Iwamoto, Y. (1969). Factors influencing productivity of cultures of *Oncomelania hupensis nosophora* (Prosobranchia: Hydrobiidae). *Am. J. Trop. Med. Hyg.*, 18, 629-637.

Davis, G. M. and Ruff, M. D. (1973). *Oncomelania hupensis* (Gastropoda: Hydrobiidae) hybridization, genetics and transmission of *Schistosoma japonicum*. *Malacol. Rev.*, 6, 181-197.

Day, P. R. (1974). "Genetics of Host Parasite Interaction." W. H. Freeman, San Francisco.

De bach, P. (1974). "Biological Control by Natural Enemies." Cambridge Univ. Press, London.

Donges, J. (1974). A formula for the mean infection success per miracidium and a method of proving the homogenous susceptibility of snail populations to trematode infection. *Int. J. Parasitol.*, 4, 403-407.

Duke, B. O. L. and Moore, P. J. (1976a). The use of a molluscicide in conjunction with chemotherapy to control *Schistosoma haematobium* at the Barombi Lake foci in Cameroon. I. The attack on the snail hosts using N-tritylmorpholine, and its effects on transmission from snail to man. *Tropenmed. Parasitol.*, 27, 297-313.

Duke, B. O. L. and Moore, P. J. (1976b). The use of a molluscicide in conjunction with chemotherapy to control *Schistosoma haematobium* at the Barombi Lake foci in Cameroon. II. Urinary examination methods, the use of Niridazole to attack the parasite in man, and the effect of transmission from man to snail. *Tropenmed. Parasitol.*, 27, 489-504.

El-Hassan, A. A. (1974). Laboratory studies of the direct effect of temperature on *Bulinus truncatus* and *Biomphalaria alexandrina*, the snail intermediate hosts of schistosomes in Egypt. *Folia Parasitol.* (Praha), 21, 181-187.

Etges, F. J. (1963). Effect of *Schistosoma mansoni* infection upon fecundity in *Australorbis glabratus*. *J. Parasitol.*, 49 (suppl.), 26.

Files, V. S. 1951. A study of the vector-parastie relationships in *Schistosoma mansoni*. *Parasitology*, 41, 264-269.

Fine, P. E. M. (rapporteur) (1976). "Mathematical Models of Schistosomiasis." *Proceed. Workshop, Bellagio, Italty*, 9-14 May 1976. Edna McConnell Clark Foundation, New York.

Fletcher, M. (1984). Genetic control of schistosomiasis: a mathematical model. *Comp. Pathobiol.*, 8. In this volume.

Frandsen, F. (1978). Hybridization between different strains of *Schistosoma intercalatum* Fisher, 1934 from Cameroun and Zaire. *J. Helminth.*, 52, 11-22.

Frandsen, F. (1979). Studies of the relationships between *Schistosoma* and their intermediate hosts. III. *J. Helminth.*, 53, 321-348.

Fuller, G. K., Lemma, A., and Haile, T. (1979). Schistosomiasis in Omo National Park in southwest Ethiopia. *Am. J. Trop. Med. Hyg.*, 28, 526-530.

Hairston, N. G. (1962). Population ecology and epidemiological problems. *In* "Ciba Foundation Symposium on Bilharziasis" (M. O'Connor, and G. Wolstenholm, eds.) pp. 36-62. Churchill, London.

Hairston, N. G. (1965). On the mathetical analysis of schistosome populations. *Bull. WHO*, 33, 45-62.

Hairston, N. G. (1973). The dynamics of transmission. *In* "Epidemiology and Control of Schistosomiasis (Bilharziasis)." (H. Ansari, ed.) pp. 250-336. University Park Press, Baltimore.

Hairston, N. G., Wrzinger, K. -H., and Burch, J. B. (1975). Non-chemical methods of snail control. World Health Organization. WHO/VBC/75.573; WHO/SCHISTO/75.40. 30 p.

Harris, K. R. (1975). The fine structure of encapuslation in *Biomphalaria glabrata*. *In* "Pathology of Invertebrate Vectors of Disease." (L. A. Bulla and T. C. Cheng eds.). *Ann. N.Y. Acad. Sci.*, 266, 446-464.

Harrison, G. 1978. "Mosquitoes, Malaria and Man." E. P. Dutton, New York.

Hoffman, D. B., Lehman, J. H., Scott, V. C., Warren, K. S., and Webbe, G. (1979). Control of schistosomiasis. *Am. J. Trop. Med. Hyg.*, 28, 249-259.

Hubendick, B. (1958). A possible method of schistosome-vector control by competition between resistant and susceptible strains. *Bull. WHO*, 18, 1113-1116.

Jelnes, J. E. (1977). Evidence of possible molluscicide resistance in *Schistosoma* intermediate hosts from Iran. *Trans. R. Soc. Trop. Med. Hyg.*, 71, 451.

Jobin, W. R., Brown, R. A., Valez, S. P., and Ferguson, F. F. (1977). Biological control of *Biomphalaria glabrata* in major reservoirs of Puerto Rico. *Am. J. Trop. Med. Hyg.*, 26, 1018-1024.

Jobin, W. R. and Laracuente, A. (1979). Biological control of schistosome transmission in flowing water habitats. *Am. J. Trop. Med. Hyg.*, 28, 916-197.

Jordan, P. (1977). Schistosomiasis - research to control. *Am. J. Trop. Med. Hyg.*, 26, 877-886.

Jordan, P. and Webbe, G. (1969). "Human Schistosomiasis." C. Thomas, Springfield, Illinois.

Kagan, I. G. and Geiger, S. (1965). The susceptibility of three strains of *Australorbis glabratus* to *Schistosoma mansoni* from Brazil and Puerto Rico. *J. Parasitol.*, 51, 622-627.

Kuris, A. M. (1973). Biological control, implications of the analogy between the trophic interactions of insect pest-parasitoid and snail-trematode systems. *Exp. Parasitol.*, 33, 365-379.

Laracuente, A., Brown, R. A., and Jobin, W. (1979). Comparison of four species of snails as potential decoys to intercept schistosome miracidia. *Am. J. Trop. Med. Hyg.*, 28, 99-105.

Liang, Y. -S. (1974). Cultivation of *Bulinus* (*Physopsis*) *globosus* (Morelet) and *Biomphalaria pfeifferi pfeifferi* (Kruass), snail hosts of schistosomiasis. *Sterkiana*, 53, 1-75.

Lie, K. J., Heyneman, D., and Richards, C. S. (1979). Specificity of natural resistance to trematode infections in *Biomphalaria glabrata*. *Int. J. Parasitol.*, 9, 529-531.

Lo, C. T. (1972). Compatibility and host-parasite relationship between species of the genus *Bulinus* (Basommatophora: Planorbidae) and an Egyptian strain of *Schistosoma haematobium* (Trematoda: Digenea). *Malacologia*, 11, 225-280.

Loker, E. S. (1983). A comparative study of the life-histories of mammalian schistosomes. *Parasitology*, 87, 343-369.

Loverde, P. T. (1976). Host-Parasite Interrelationships Between the Trematode *Schistosoma haematobium* from Egypt and Polyploid Snails of the Genus *Bulinus*. Ph.D. Dissertation, Univ. Michigan, Ann Arbor, Michigan.

MacDonald G. (1965). The dynamics of helminth infections with special reference to schistosomes. *Trans. R. Soc. Trop. Med. Hyg.*, 59, 489-506.

MacDonald, G. (1973). Measurement of the clinical manifestations of schistosomiasis. *In* "Epidemiology and Control of Schistosomiasis." (N. Ansari, ed.). pp. 354-387. University Park Press, Baltimore, Maryland.

MacDonald, W. W. (1976). Mosquito genetics in relation to filarial infections. *Symp. Br. Soc. Parasitol.*, 14, 1-24.

Malek, E. A. (1978). Realistic goals in the use of molluscicides in different endemic areas of schistosomiasis. *Proc. Intl. Conf. Schistosomiasis, Cairo, Egypt, October* 18-25, 1975, 1 359-391.

Malek, E. A.   (1980).   "Snail-Transmitted Parasitic Diseases."
    Vol. I.   CRC Press, Boca Raton, Florida.

Mansour, N. S.   (1973).   *Schistosoma mansoni* and *Schistosoma
    haematobium* natural infection in the Nile rat *Arvicanthis n.
    nicoticus* from an endemic area in Egypt.   *J. Egypt. Public.
    Health Assoc.*, 48, 94-100.

Markel, S. F., Loverde, P. T., and Birtt, E. M.   (1979).   Prolonged
    latent schistosomiasis.   *J. Am. Med. Assoc.*, 240, 1746-1747.

McClelland, W.F.J.   (1965).   Development of *Schistosoma haematobium*
    in *Bulinus (Physopsis) nasutus*.   *E. Afr. Inst. Med. Res. Ann.
    Rep.*, 1963-64, Mwanza, 1965.   pp. 15-17.

Michelson, E. H. and Dubois, L.   (1978).   Susceptibility of Bahian
    populations of *Biomphalaria glabrata* to an allopatric strain
    of *Schistosoma mansoni*.   *Am. J. Trop. Med. Hyg.*, 27, 782-786.

Minchella, D. J.   (1983).   Laboratory comparison of the relative
    success of *Biomphalaria glabrata* stocks which are susceptible
    and insusceptible to infection with *Schistosoma mansoni*.
    Parasitology, 86, 335-344.

Morais, J.A.D.   (1975).   Schistosomiase mansoni em Angola: notas
    sobre a sua recente difusao.   *An. Inst. Hig. Med. Trop.*, 3,
    405-423.

Mulvey, M. and Vrijenhoek, R. C.   (1981).   Multiple paternity in
    the hermaphroditic snail *Biomphalaria obstructa*.   *J. Hered.*,
    72, 308-312.

Mulvey, M. and Vrijenhoek, R. C.   (1982).   Population structure in
    Biomphalaria glabrata: examination of an hypothesis for the
    patchy distribution of susceptibility to schistosomes.   *Am.
    J. Trop. Med. Hyg.*, 31, 1195-1200.

Newton, W. L.   (1952).   The comparative tissue reaction of two
    strains of *Australorbis glabratus* to infection with
    Schistosoma mansoni.   *J. Parasitol.*, *38, 362-366.*

Newton, W. L.   (1953).   The inheritance of susceptibility to infection
    with *Schistosoma mansoni* in *Australorbis glabratus*.   *Exp.
    Parasitol.*, 2, 242-257.

Newton, W. L.   (1955).   The establishment of a strain of *Australorbis
    glabratus* which combines albinism and high susceptibility to
    infection with *Schistosoma mansoni*.   *J. Parasitol.*, 41, 526-
    528.

Otori, Y. Ritchie, L. S., and Hunter, G. W. (1956). The incubation period of the egg of *Oncomelania nosophora*. *Am. J. Trop. Med. Hyg.*, 5, 559-561.

Pal, R. and Whitten, M. J. (eds.) (1974). "The Use of Genetics in Insect Control." Elsevier, London.

Paperna, I. (1968). Susceptibility of *Bulinus* (*Physopsis*) *globosus* and *Bulinus truncatus rohlfsi* from different localities in Ghana to different local strains of *Schistosoma haematobium*. *Ann. Trop. Med. Parasitol.*, 62, 12-26.

Paraense, W. L. (1955). Self- and cross-fertilization in *Australorbis glabratus*. *Mem. Inst. Oswaldo Cruz, Rio de Janeiro*, 53, 285-291.

Paraense, W. L. and Correa, L. R. (1963). Variation in susceptibility of populations of *Australorbis glabratus* to a strain of *Schistosoma mansoni*. *Rev. Inst. Med. Trop. Sao Paulo*, 5, 15-22.

Paraense, W. L. and Correa, L. R. (1978). Differential susceptibility of *Biomphalaria tenagophila* populations to infection with a strain of *Schistosoma mansoni*. *J. Parasitol.*, 64, 822-826.

Pesigan, T. P., Hairston, N. G., Jauregui, J. J., Garcia, E. G., Santos, A. T., Santos, B. C., and Besa, A. A. (1958). Studies on *Schistosoma japonicum* infection in the Philippines. *Bull. WHO*, 18, 481-578.

Plapp, F. W. (1976). Biochemical genetics of insecticide resistance. *Annu. Rev. Entomol.*, 21, 179-197.

Rachford, F. W. (1977). *Oncomelania hupensis quadrasi* from Mindoro (Victoria), Leyte (Palo), and Mindanaao (Davao del Norte) of the Philippines: susceptibility to infection with Philippine isolates of *Schistosoma japonicum*. *J. Parasitol.*, 63, 1129-1130.

Richards, C. S. (1970). Genetics of a molluscan vector of schistosomiasis. *Nature*, 227, 806-810.

Richards, C. S. (1973a). Susceptibility of adult *Biomphalaria glabrata* to *Schistosoma mansoni* infection. *Am. J. Trop. Med. Hyg.*, 22, 748-756.

Richards, C. S. (1973b). Genetics of *Biomphalaria glabrata* (Gastropoda: Planorbidae). *Malacol. Rev.*, 6, 199-202.

Richards, C. S. (1975a). Genetics factors in susceptibility of *Biomphalaria glabrata* for different strains of *Schistosoma mansoni*. *Parasitology*, 70, 221-241.

Richards, C. S. (1975b). Genetic studies on variation in infectivity of *Schistosoma mansoni*. *J. Parasitol.*, 61, 233-236.

Richards, C. S. (1975c). Genetic studies of pathologic conditions and susceptibility to infection in *Biomphalaria glabrata*. *Ann. N.Y. Acad. Sci.*, 266, 394-410.

Richards, C. S. (1976). Genetics of the host-parasite relationship between *Biomphalaria glabrata* and *Schistosoma mansoni*. *Symp. Br. Soc. Parasitol.*, 14, 45-54.

Richards, C. S. (1977). *Schistosoma mansoni:* susceptibility reversal with age in the snail host *Biomphalaria glabrata*. *Exp. Parasitol.*, 42, 165-168.

Richards, C. S. (1983). Influence of snail age on genetic variations in susceptibility of *Biomphalaria glabrata* for infection with *Schistosoma mansoni*. *Malacologia*, 25, 493-502.

Richards, C. S. and Merritt, J. W. (1972). Genetic factors in the susceptibility of juvenile *Biomphalaria glabrata* to *Schistosoma mansoni* infection. *Am. J. Trop. Med. Hyg.*, 21, 425-434.

Santana, J. V. de, Magahlaes, L. A. and Rangel, H. de A. (1978). Selecao de linhagens de *Biomphalaria tenagophila* e *Biomphalaria glabrata* visando maior suscetibilidade ao *Schistosoma mansoni*. *Rev. Saude Publ., Sao Paulo*, 12, 67-77.

Saoud, M.F.A. 1965. Susceptibilities of various snail intermediate hosts of *Schistosoma mansoni* to various strains of the parasite. *J. Helminth.*, 39, 363-376.

Southgate, V. R., Van Wijk, H. B., and Wright, C. A. (1976). Schistosomiasis at Loum, Cameroun; *Schistosoma haematobium, S. intercalatum* and their natural hybrid. *Z. Parasitenk.*, 49, 145-159.

Stunkard, H. W. (1946). Possible snail host of human schistosomes in the United States. *J. Parasitol.*, 32, 539-552.

Sturrock, B. M. (1966). The influence of infection with *Schistosoma mansoni* on the growth rate and reproduction of *Biomphalaria pfeifferi*. *Ann. Trop. Med. Parasitol.*, 60, 187-197.

Sturrock, R. F. (1973). Field studies on the transmission of *Schistosoma mansoni* and on the bionomics of its intermediate host, *Biomphalaria glabrata*, on St. Lucia, West Indies. *Int. J. Parasitol.*, 3, 175-194.

Sudds, R. H. (1960). Observations of schistosome miracidial behavior in the presence of normal and abnormal snail hosts and subsequent tissue studies of the hosts. *J. Elisha Mitchell Sci. Soc.*, 76, 121-123.

Sullivan, J. T., Cheng, T. C., and Chen, C. C. (1984). Genetic selection for tolerance to niclosamide and copper in *Biomphalaria glabrata* (Mollusca: Pulmonta). *Tropenmed. Parasit.*, 35, 189-192.

Theron, A., Pointier, J.-P., and Combes, C. (1978). Approche ecologique du probleme de la responsabilite de l'homme et du rat dans le functionement d un site de transmission a *Schistosoma mansoni* en Guadeloupe. *Ann. Parasitol. (Paris)*, 53, 223-234.

Thomas, J. D. (1973). Schistosomiasis and the control of molluscan hosts of human schistosomes with particular reference to possible self-regulatory mechanisms. *Adv. Parasitol.*, 11, 307-338.

Uhazy, L. S., Tanaka, R. D., and MacInnis, A. J. (1978). *Schistosoma mansoni:* identification of chemicals that attract or trap its snail vector, *Biomphalaria glabrata*. *Science*, 201, 924-926.

Upatham, E. S. (1972). Interference by unsusceptible aquatic animals with the capacity of the miracidia of *Schistosoma mansoni* Sambon to infect *Biomphalaria glabrata* (Say) under field-simulated conditions in St. Lucia, West Indies. *J. Helminth.*, 66, 277-283.

Upatham, E. S. and Sturrock, R. F. (1973). Field investigations on the effect of other aquatic animals on the infection of *Biomphalaria glabrata* by *Schistosoma mansoni* miracidia. *J. Parasitol.*, 59, 448-453.

Van Den Bosch, R. and Messenger, P. S. (1973). "Biological Control." Intext, New York.

Wakelin, D. (1978). Genetic control of susceptibility and resistance to parasitic infections. *Adv. Parasitol.*, 16, 219-308.

Walton, B. C., Winn, M. M., and Williams, J. E. (1958). Development of resistance to molluscicides in *Oncomelania nosophora*. *Am. J. Trop. Med. Hyg.*, 7, 618-619.

Warren, K. S., Mahmoud, A.A.F., Cummings, P., Murphy, D. J., and Houser, H. B. (1974). *Schistosomiasis mansoni* in Yemeni in California: duration of infection, presence of disease, therapeutic management. *Am. J. Trop. Med. Hyg.*, 23, 902-909.

Webbe, G. (1978). Epidemiology of schistosomiasis and prospects for control. *Proc. Int. Conf. Schistosomiasis, Cairo, Egypt, October* 18-25, 1975, 1, 13-22.

Woodruff, D. A. (1978). Biological control of schistosomiasis by genetic manipulation of intermediate-host snail populations. *Proc. Int. Conf. Schistosomiasis, Cairo, Egypt,* October 18-25, 1975, 2, 755.

Woodruff, D. S. (1983). An approach to epidemiology. *Science*, 222, 1321-1322.

World Health Organization. (1954). Study group on bilharzia snail vector identification and classification. *WHO Tech. Rep. Ser.* No. 90.

Wright, C. A. (1968). Some views on the biological control of trematode diseases. *Trans. R. Soc. Trop. Med. Hyg.*, 62, 320-324.

Wright, C. A. (1971a). Review of "Genetics of a Molluscan Vector of Schistosomiasis" by C. S. Richards. *Trop. Dis. Bull.*, 68, 333-335.

Wright, C. A. (1971b). "Flukes and Snails." Allen and Unwin, London.

Wright, C. A. (1974). Snail susceptibility or trematode infectivity? *J. Nat. Hist.*, 8, 545-548.

Wright, C. A. and Southgate, V. R. (1976). Hybridization of schistosomes and some of its implications. *Symp. Br. Soc. Parasitol.*, 14, 55-86.

Wright, C. A. and Southgate, V. R. (1981). Coevolution of digeneans and molluscs, with special reference to schistosomes and their intermediate hosts. *In* "The Evolving Biosphere" (P. L. Forey, ed.). pp. 191-205. Cambridge University Press, Cambridge, England.

Wright, C. A., Southgate, V. R., Van Wijk, H. B., and Moore, P. J.
    (1974). Hybrids between *Schistosoma haematobium* and *S.
    intercalatum* in Cameroun. *Trans. Roy. Soc. Trop. Med. Hyg.*,
    68, 413-414.

Wright, W. H. (1973). Geographical distribution of schistosomes
    and their intermediate hosts. *In* "Epidemiology and Control
    of Schistosomiasis (Bilharziasis)." (N. Ansari, ed.).
    pp. 32-249. University Park Press, Baltimore, Maryland.

Yasuraoko, K. (1972). Studies of the resistance of *Oncomelania*
    snails to molluscicides. *In* "Research in Filariasis and
    Schistosomiasis in Japan." (M. Yokogawa, ed.). pp. 103-111.
    University Park Press, Baltimore, Maryland.

GENETIC CONTROL OF SCHISTOSOMIASIS: A MATHEMATICAL MODEL

Madeleine Fletcher[1]

Department of Epidemiology and Public Health
Yale University School of Medicine
New Haven, Connecticut 06510

[1]Department of Community Health, University of Addis, Abeba, Ethiopia.

I. INTRODUCTION

Genetic manipulation of the intermediate host snails of human-infecting schistosomes has recently attracted interest as a potential method for control of schistosomiasis (Richards, 1970; Woodruff, 1978, 1985). Here I develop a simple mathematical model to illustrate the principles underlying this form of genetic control and to serve as a springboard for a discussion of the effectiveness of this approach.

The proposed method is based on evidence that snail-schistosome compatibility is genetically controlled in both organisms (reviewed by Woodruff, in this volume). Snails can be selected for resistance to the schistosome strain they normally transmit. The control method would involve increasing the resistance of snail populations by re-peated releases of conspecific but resistant snails. Resistant snails may have a selective advantage over susceptible parasitized snails, which suffer increased mortality and decreased fertility (reviewed by Meuleman, 1972; Anderson and May, 1979). This selec-tive advantage would contribute to the spread of genes for resis-tance through the snail population. The resistant snails would also act as decoys to intercept schistosome miracidia. In this last respect only, genetic control is similar in approach to the recently proposed snail decoy technique (reviewed by Laracuente et al., 1979), which involves the introduction of nonsusceptible species of snails rather than of conspecific resistant snails. The basic assumption underlying the idea of genetic control dis-cussed here is that a significant increase in the resistance to schistosome infection of snail populations and concurrent decrease in the snail infection rate will lead to an effective decline in transmission of the disease.

One of the advantages of this method of snail control lies in the fact that it is based on the principle of population replace-ment, rather than that of eradication. Snail eradication using molluscicides is very difficult to achieve because of the snails' high reproductive capacity and short generation time. Genetic control on the other hand makes good use of these same characteris-tics. Genetic control is ecologically sound, as opposed to chemical control, and may present some advantages over biological control, which involves the introduction of foreign species with typically unpredicatable consequences. Genetic control is highly specific, and is generally compatible with other control methods. The outcome of computer simulations generated by the present mathematical model suggests that genetic manipulation of the intermediate host snails of schistosomes should be given serious consideration as an auxiliary method of control of the disease.

II. A MATHEMATICAL MODEL OF GENETIC CONTROL

The present model, based on the *Schistosoma mansoni-Biomphalaria* association, explores the manner in which the prevalence of snail infections is affected by the introduction of resistant snails into a population. The model thus does not cover the entire schistosome life cycle, but rather focuses on the larval intramolluscan phase. Existing models of the entire transmission process (reviewed by Cohen, 1977; May and Anderson, 1979) are of necessity very general and ignore many details, including snail and parasite genetic heterogeneity. It is not yet possible to develop a comprehensive model integrating all significant aspects of the schistosome life cycle because the manner in which various parameters regulate transmission is still unclear and accurate estimates of many of these parameters are lacking (for examples, see Barbour, 1978, 1982; Anderson and May, 1979). Such a comprehensive model could eventually be used to predict the rate of decline of adult schistosome numbers in response to increased snail resistance, this of course being the ultimate criterion by which to judge the effectiveness of genetic control.

Assumptions

The model is designed to illustrate general principles and mechanisms underlying the theory of genetic control, rather than to quantitate accurately what will happen in any given situation. Many simplifying assumptions are thus made in order to delimit precisely the subject of investigation.

For instance, the simplest genetic system of snail-schistosome compatibility is considered, in which the parasite is genetically uniform with respect to infectivity, and snail susceptibility is determined by one gene, with one allele conferring resistance and the other susceptibility. (Two situations where age-specific susceptibility appears to be monogenic are documented by Richards, 1970, 1975a). Two cases are analyzed, one where the allele for resistance is dominant and the other where it is recessive. In this model, susceptibility confers a selective disadvantage, based on increased mortality and decreased fertility of infected snails. The snails presumably mate randomly with respect to the alleles regulating susceptibility, so that gene frequencies change in accordance with Hardy-Weinberg expectations for large out-breeding populations [a valid assumption because even thought *Biomphalaria* are facultative hermaphrodites, they preferentially cross-fertilize (Paraense, 1955; Mulvey and Vrijenhoek, 1981, 1982; Woodruff, Mulvey and Yipp, unpubl.]. Other simplifying assumptions on the genetic structure of the snail population include nonoverlapping generations and negligible mutation and migration rates.

Miracidia are assumed to have equal penetration success in resistant and susceptible snails. Evidence that miracidia do not discriminate between resistant and susceptible snail species is reviewed by Chernin (1970). A successful infection occurs when a miracidium penetrates a susceptible snail and is able to complete development in it. An infection is not successful when the penetrated snail is resistant, or when development is not completed in a susceptible snail because of nongenetic factors, such as miracidial age or parasite crowding.

In this deterministic model, the only variable involved in determining the snail infection rate is the proportion of genetically susceptible snails, all other factors remaining constant. In particular, population sizes of snails and of adult schistosomes are assumed to remain constant, seasonal and chance fluctuations being ignored.

Snail Infection Rate

The snail infection rate (I, a variable) is the proportion of snails penetrated and successfully infected in the total population. Since it is assumed that the only variable involved in determining I is the proportion of susceptible snails, C, the infection rate in a totally susceptible snail population is a constant, k. This constant k is the thus a quantitative index of the intensity of schistosome transmission under a specific set of conditions, and incorporates miracidial density and all other nongenetic factors, environmental and biological, contributing to infection success. Readers interested in a mathematical treatment of the relationship between the snail infection rate and these nongenetic factors are referred to Anderson (1978). Given the assumptions that miracidia are genetically homogeneous with regard to infectivity and have equal penetration success in resistant and susceptible snails, I is directly proportional to C, so that:

$$I = k.C \qquad\qquad (1)$$

This formula for the snail infection rate is specific for the particular genetic system considered (two phenotypes for snail susceptibility and one for miracidial infectivity). For a more generalized mathematical treatment of the relationship between the probability of a snail infection and genetic factors in both snail and parasite, Basch (1975) should be consulted. The formula he develops can be adapted to any number of phenotypes involved in determining compatibility.

Phenotype Fitness

Fitness is defined in this model as the proportional contribution of offspring to the next generation (following Falconer, 1960). The fitness of a particular phenotype is a product of both genetic and environmental factors which vary from individual, and so is calculated here as an average value.

In this model the difference between the fitness of susceptible and resistant snails is based on the burden of infection on the susceptible snails, and is a density-independent factor. It is thus a function of k and of the difference in reproductive rates between infected and uninfected snails (the reproductive rate accounts for differences in mortality as well as in fertility). Resistant snails are assumed to have the fitness of uninfected susceptible snails.

Let R_U be the reproductive rate in uninfected snails, both susceptible and resistant, and let R_I be the reproductive rate of infected snails. The average reproductive rate of susceptible snails, R_S, in a population with an infection rate I and a susceptible proportion C, is thus:

$$R_S = R_U \cdot (C - I)/C + R_I \cdot I/C \qquad (2)$$

Setting the fitness of resistant snails, R_U, as 1.0, the relative fitness of susceptible snails, W is:

$$W = R_S/R_U = 1 - I/C + R_I/R_U \cdot I/C$$

$$= 1 - k \cdot (1 - R_I/R_U) \qquad (3)$$

Because k, R_U, and R_I are held constant, the relative fitness of susceptible snails also remains constant throughout each simulation.

Change in Gene Frequency

Let p be the frequency of the resistance allele, and q the frequency of the susceptibility allele (p + q = 1). We assume the original snail population to be composed entirely of homozygous susceptible individuals. with an initial infection rate k (from equation 1). If the resistance allele is introduced into the snail population at a ratio b of the number of susceptibility alleles, its frequency p, which is then b/(1 + b), will slowly begin to increase, as a consequency of hybridization between susceptible snails and resistant snails of assumed superior fitness. The equations (4-6) formalizing the increase in frequency of p over

generations are developed in Table 1, for the case in which resistance is dominant, and in Table 2, for the case in which is is
recessive. As the frequency of the resistance allele increases,
the snail infection rate I, which is proportional to the number of
susceptibility snails (equation 1) and thus to the frequency of
the susceptible allele, decreases.

Computer Simulations

A computer program based on the equations for snail infection
rate, relative fitness, and gene frequency of the resistance allele
(equations 1 through 6) was developed. Various simulations
monitored the rate of increase of the resistant phenotype and the
rate of decrease of the snail infection rate, in response to single
or repeated releases of resistant snails. The reproductive rates
R_U and R_I were calculated using the formula $\Sigma\ l_x.m_x;l_x$ being the
age-specific mortality, m_x the age-specific fertility, taking
account of sterile or unhatched eggs (Slobodkin, 1961). From
laboratory data published by Sturrock (1966) on *Biomphalaria
pfeifferi* from Tanzania infected with *S. mansoni*, I calculated
R_U to be 1693.6, R_I to be 26.8. This reflects almost total inhibition of egg production in infected snails. Three different levels
of k, the initial snail infections rate, were considered, 0.01,
0.10, and 0.25. The relative fitnesses of the susceptible phenotype
corresponding to these initial infection rates were 0.990, 0.092,
and 0.754, respectively. The effect of introducing a resistant
phenotype of inferior fitness, because of loss of field adapt-
ability, for instance, was also simulated. The ratio of resistant
snails released in relation to the number of susceptible snails
varied from 1:10 to 10:1. The frequency of release varied from a
single initial release to one release per snail generation. The
snail generation time was assumed to be 3 months [generation times
as low as 1 month have been reported (Ritchie et al., 1963)]. The
rate of increase in frequency of a mutant resistance gene under
natural selection was simulated by a single initial release of
resistant snails at a 1:10,000 ratio.

III. RESULTS

Figures 1 and 2 compare the effect of genetic control on snail
infection rates at three different levels of schistosome transmission, with resistance a dominant trait, and resistance a
recessive trait, respectively. These figures show that the decrease
in the prevalence of snail infections in response to yearly releases of resistant snails in a 1:10 ratio occurs significantly
faster when the initial infection rate k is high. This is because
the selective pressure against susceptible snails increases with
k (equation 4). However, the influence of k on the rate of decrease in prevalence becomes insignificant when more frequent

TABLE 1. Increase in frequency of the resistance allele in a snail population when resistance to schistosomes is a dominant trait.*

GENOTYPES	RR	Rr	rr	TOTALS
FITNESS	1	1	$1-s$	
Initial gene frequencies+	$p_0 = \dfrac{b}{1+b}$	0	$q_0 = \dfrac{1}{1+b}$	1
Gametic contribution to F_1 (after selection)	p_0	0	$q_0 \cdot (1-s)$	$1 - s \cdot q_0$
Gene frequency in F_1	$p_1 = \dfrac{p_0}{1 - s \cdot q_0}$ (4)		$q_1 = \dfrac{q_0 \cdot (1-s)}{1 - s \cdot q_0}$	1
Genotype frequencies in F_1	p_1^2	$2 \cdot p_1 \cdot q_1$	q_1^2	1
Gametic contribution to F_2 (after selection)	p_1^2	$2 \cdot p_1 \cdot q_1$	$q_1^2 \cdot (1-s)$	$1 - s \cdot q_1^2$‡

TABLE 1. (Con't.)

GENOTYPES FITNESS	RR 1	Rr 1	rr $1 - s$	TOTALS
Gene frequency in F_2 §	$p_2 = \dfrac{p_1}{1 - s \cdot q_1^2}$		$q_2 = \dfrac{q_1 \cdot (1 - s \cdot q_1)}{1 - s \cdot q_1^2}$	1
Gene frequency in F_n 11	$p_n = \dfrac{p_{n-1}}{1 - s \cdot q_{n-1}^2}$ (5)			

* R represents the resistance allele, with frequency p, r the susceptibility allele, with frequency q (p + q = 1). The fitness of the resistant phenotype (RR and Rr) is set at 1, the relative fitness of the susceptible phenotype (rr) is represented by 1 − s (s is a constant whose value is expressed by k.(1 − R_I/R_{IJ}), see text). F_1, F_2...F_n represent the generation number.

+ at time of initial release of resistant snails at a ratio b of the number of susceptible snails.

+

+ derived by adding the 3 post-selection frequencies and equating $p^2 + 2 \cdot p \cdot q + q^2 = 1$.

§ derived by adding the RR or rr frequency to half that of the Rr frequency.

11 by recurrence.

TABLE 2. Increase in frequency of the resistance allele in a snail population when resistance to schistosomes is a recessive trait.*

Genotypes Fitness	$s^r s^r$ 1	Ss^r 1-s	SS 1-s	TOTALS
Initial gene frequencies[+]	$p_o = \dfrac{b}{1+b}$	0	$q_o = \dfrac{1}{1+b}$	1
Gametic contributions to F_1 (after selection)	p_o	0	$q_o \cdot (1-s)$	$1 - s \cdot q_o$
Gene frequency in F_1	$p_1 = \dfrac{p_o}{1 - s \cdot q_o}$		$q_1 = \dfrac{q_o \cdot (1-s)}{1 - s \cdot q_o}$	1
Genotype Frequencies in F_1	p_1^2	$2 \cdot p_1 \cdot q_1$	q_1^2	1

TABLE 2. (Con't).

Genotypes Fitness	$s^r s^r$ 1	Ss^r 1-s	SS 1-s	TOTALS
Gametic contribution to F_2 (after selection)‡	p_1^2	$2 \cdot p_1 \cdot q_1 (1-s)$	$q_1^2 \cdot (1-s)$	$1 - s + s \cdot p_1^2$
Gene frequency in F_2 §	$p_2 = \dfrac{p_1 \cdot (1 - s + s \cdot p_1)}{1 - s + s \cdot p_1^2}$		$q_2 = \dfrac{q_1 \cdot (1-s)}{1-s+s \cdot p_1^2}$	1
Gene frequency in F_n 11	$p_n = \dfrac{p_{n-1} \cdot (1-s+s \cdot p_{n-1})}{1 - s + s \cdot p_{n-1}^2}$ (6)			

* r represents the resistance allele, with frequency p, S the susceptibility allele with frequency q (p + q = 1). The fitness of the resistant phenotype ($s^r s^r$) is set at 1, the relative fitness of the susceptible phenotype (SS and Ssr) is represented by 1 - s (s as in Table 1). F_1, F_2, ...F_n, and b as in Table 1.

+ at time of initial release of resistant snails.

$^{+}$

$^{+}$derived as in Table 1.

^{s}derived by adding the $s^{r}s^{r}$ or SS frequency to half that of the Ss^{r} frequency and equating $p = 1 - q$ and $q = 1 - p$.

^{11}by recurrence.

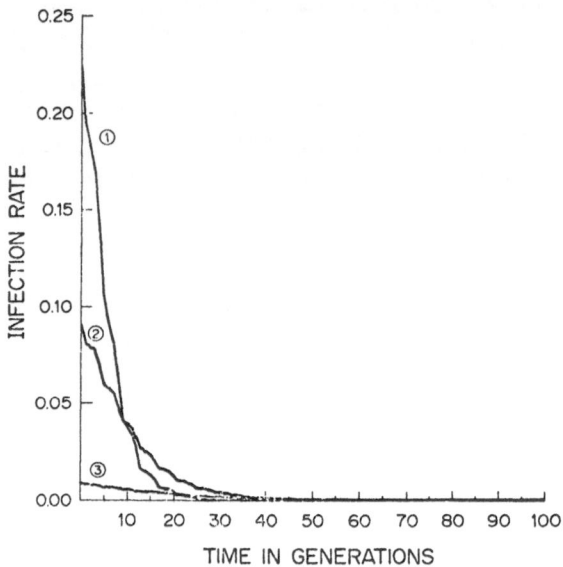

FIGURE 1.   Change in the snail infection rate over time measured in
            snail generations, in response to releases of resistant
            snails every four generations at a 1:10 ratio, at 3
            different levels of schistosome transmission; 1. k = 0.25.
            2. k = 0.10. 3. k = 0.01; resistance is a dominant
            trait; snails are released at time 0.

releases at a higher ratio are used, such as 1:1 twice yearly
(see Tables 3 and 4). Comparing Figure 1 to Figure 2 it can be seen
that the snail infection rate decreases more slowly when resistance
is recessive than when it is dominant (also see Tables 3 and 4).

    Figures 3 and 4 illustrate the corresponding increase in the
proportion of resistant snails in response to yearly releases at a
1:10 ratio, at the three different levels of transmission, with
resistance dominant, and resistance recessive, respectively. With
resistance a recessive trait (Fig. 4) and k = 0.01, each increase
in the proportion of resistant snails resulting from a release is
followed by a smaller temporary decrease, caused by hybridization
of the incoming snails with susceptible wild snails and the subsequent production of heterozygous susceptible progeny. This explains
why the proportion of resistant snails increases more slowly when
resistance is recessive.

    Figures 5 and 6 contrast the effect on the snail infection rate
of a single release of resistant snails at the outset of control,

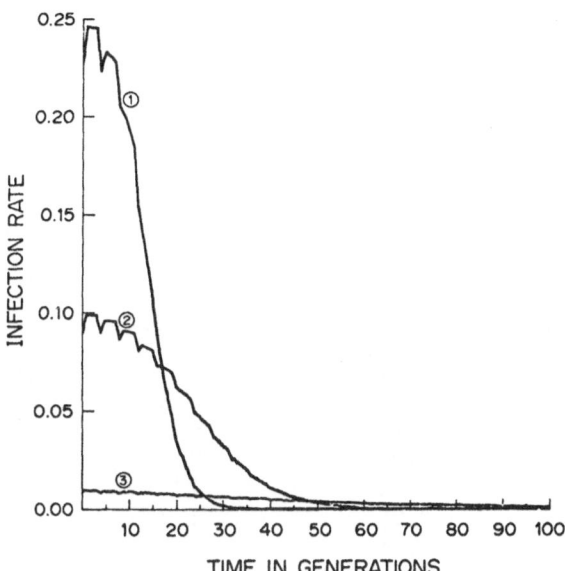

FIGURE 2. Change in the snail infection rate over time measured
in snail generations, in response to releases of resistant snails every 4 generations at a 1:10 ratio,
at 3 different levels of schistosome transmission;
1. k = 0.25. 2. k = 0.10. 3. k = 0.01; resistance is a
recessive trait; snails are released at time 0.

TABLE 3. Summary of results for the case in which resistance is
dominant.*

Frequency of Release	Release Ratio	Initial Infection Rate k	Time to $I = 10^{-2}$	Time to $I = 10^{-4}$	Time to $I = 10^{-6}$
Single Release	1:10,000	0.01	0	> 25	> 25
		0.10	> 25	> 25	> 25
		0.25	13	> 25	> 25
	1:10	0.01	0	> 25	> 25
		0.10	12.25	> 25	> 25
		0.25	7	> 25	> 25

TABLE 3. (Con't).

Frequency of Release	Release Ratio	Initial Infection Rate k	Time to $I = 10^{-2}$	Time to $I = 10^{-4}$	Time to $I = 10^{-6}$
Single Release	10:1	0.01	0	0.25	> 25
		0.10	0	> 25	> 25
		0.25	0.25	> 25	> 25
Once a Year	1:10	0.01	0	21	> 25
		0.10	5.5	16.25	> 25
		0.25	4	9.75	15.75
	1:1	0.01	0	3	6.25
		0.10	1	4	7
		0.25	1.25	3.25	6
	10:1	0.01	0	0.25	1.25
		0.10	0	1	2
		0.25	0.25	1	2
Twice a Year	1:10	0.01	0	11	21.75
		0.10	3.5	9.25	15.25
		0.25	2.75	5.75	9
	1:1	0.01	0	1.50	3
		0.10	0.5	2	3.5
		0.25	0.75	2	3.25
	10:1	0.01	0	0.25	0.75
		0.10	0	0.5	1
		0.25	0.25	0.75	1
Four Times a Year	1:10	0.01	0	5.75	11
		0.10	2	5.25	8.25
		0.25	1.75	3.5	5
	1:1	0.01	0	0.75	1.75
		0.10	0.5	1	1.75
		0.25	0.5	1	1.75
	10:1	0.01	0	0.25	0.5
		0.10	0	0.15	0.5
		0.25	0.25	0.5	0.5

*Time units are years (4 snail generations per year); I is the snail infection rate, initially equal to k.

TABLE 4. Summary of results for the case in which resistance is recessive.*

Frequency of Release	Release Ratio	Initial Infection Rate k	Time to $I = 10^{-2}$	Time to $I = 10^{-4}$	Time to $I = 10^{-6}$
Single Release	1:10,000	0.01	0	>250	>250
		0.10	>250	>250	>250
		0.25	>250	>250	>250
	1:10	0.01	0	> 25	> 25
		0.10	> 25	> 25	> 25
		0.25	11.5	15.75	19.75
	10:1	0.01	0	> 25	> 25
		0.10	1.5	13	24
		0.25	1.5	5.75	9.75
Once a Year	1:10	0.01	0	> 25	> 25
		0.10	10.25	19.75	> 25
		0.25	6.25	10	13.75
	1:1	0.01	0	7	13
		0.10	3	7	11
		0.25	2.5	5	7.5
	10:1	0.01	0	2	4
		0.10	0	2	4
		0.25	1	2	3.25
Twice a Year	1:10	0.01	0	23.75	> 25
		0.10	7	15	22.5
		0.25	4.75	8.25	11.75
	1:1	0.01	0	3.5	6.5
		0.10	1.5	4	6.5
		0.25	1.5	3.5	5.5
	10:1	0.01	0	1	2
		0.10	0.5	1	2
		0.25	0.5	1.25	2

TABLE 4. (Con't).

Frequency of Release	Release Ratio	Initial Infection Rate k	Time to $I = 10^{-2}$	Time to $I = 10^{-4}$	Time to $I = 10^{-6}$
Four Times A Year	1:10	0.01	0	12.75	23.5
		0.10	4.5	10.5	16.25
		0.25	3.5	6.5	9.5
	1:1	0.01	0	1.75	3.5
		0.10	0.75	2.25	3.75
		0.25	1	2.25	3.25
	10:1	0.01	0	0.5	1
		0.10	0	0.75	1
		0.25	0.25	0.75	1

*Time units I and k as in Table 3.

versus repeated releases, once, twice, and four times yearly. Whether resistance is dominant (Fig. 5) or recessive (Fig. 6), it is evident that a single release at a 1:10 ratio results in a slow rate of change when k is 0.1. Tables 3 and 4 show this is the case at all three levels of transmission considered, even at release ratios of up to 10:1.

If one snail in 10,000 is resistant as a consequence of mutation, and natural selection for the resistant phenotype is allowed to proceed, it would take over 25 years in most cases for the snail infection rate to drop to 0.01 when resistance is dominant, and over 250 years when it is recessive.

IV. DISCUSSION

The model developed above illustrates several interesting points relevant to the development of an efficient strategy of genetic control.

(1) The model shows that in nature, the rise in frequency of a resistance gene would be very gradual. There would be ample time for the spread of resistance through the snail population to result in selection in favor of variant schistosomes capable of parasitizing the resistant snails. Snail populations would remain susceptible to infection in nature simply because parasites coevolve along with their hosts. It is thus not necessary to hypothesize the existence of a selective disadvantage associated with resistance to explain

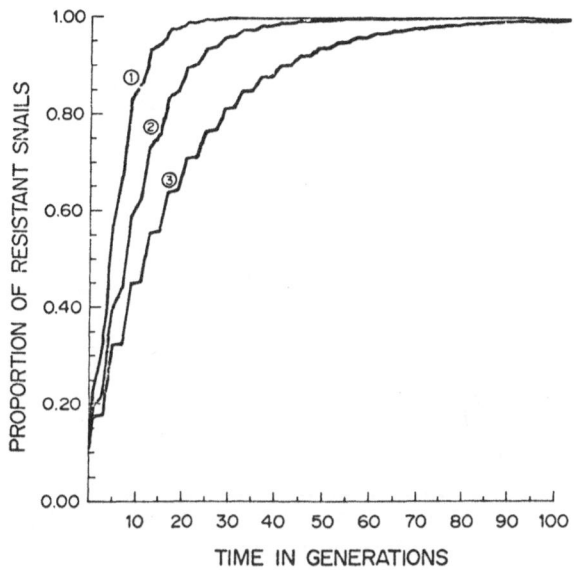

FIGURE 3.   Change in the proportion of resistant snails in the popu-
            lation over time measured in snail generations, in response
            to releases of resistant snails every 4 generations at a 1:10
            ratio, at 3 different levels of schistosome transmission;
            1. k = 0.25. 2. k = 0.10. 3. k = 0.01; resistance is a
            dominant trait; snails are released at time 0.

why snail populations have not evolved total resistance in nature
(e.g., Wright, 1971; Fine, 1976).

      (2) The model shows that with a single release of resistant
snails the spread of resistance through the snail population is
relatively slow. Because natural selection acts to maintain snail
and schistosome populations in genetic balance, a drastic and sus-
tained perturbation is necessary to disrupt that balance and lead
to a rapid rise in the frequency to disrupt that balance and lead
to a rapid rise in the frequency of resistance genes before the
parasite can adapt. The model shows that although a single release
is relatively ineffective, repeated releases over a period of 2 to
4 years can bring about a rapid increase in the resistance of the
snail populations.

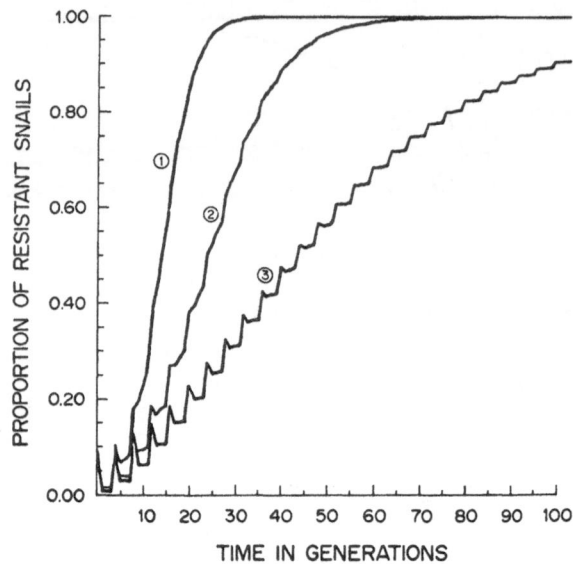

FIGURE 4. Change in the proportion of resistant snails in the
population over time measured in snail generations, in
response to releases of resistant snails every 4 generations at a 1:10 ratio, at 3 different levels of
schistosome transmission; 1. k = 0.25. 2. k = 0.10.
3. k = 0.01; resistance is a recessive trait; snails are
released at time 0.

(3)   Figures 1 through 6 show that the rise in numbers of
resistant snails and corresponding decline in the snail infection
rate in response to genetic control are asymptotic; although
initially the rate of change is rapid, it slows down gradually
as the population approaches complete resistance. Thus, as with
chemical snail control, in which the snail population cannot
usually be eradicated, we are led to consider the question of
effectiveness, i.e., the determination of the tolerable frequency
of infected snails consistent with disease control. In general
that frequency must be very low, especially where snail populations
are large. Again, Tables 3 and 4 emphasize the importance both of
repeated releases and of adequate release ratios in reducing
drastically the snail infection rate within a reasonable time inter-
val. For example, under the conditions of the model, releases of
resistant snails at a 1:1 ratio two to four times a year should
reduce the snail infection rate from 0.1 to 0.000001 within 2 to
4 years.

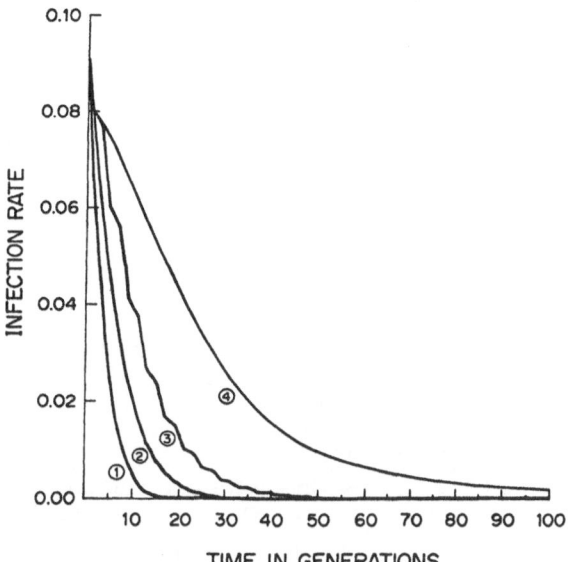

FIGURE 5.   Change in the snail infection rate over time measured in
            snail generations, in response to releases of resistant
            snails at a 1:10 ratio; 1. every generation. 2. every
            2 generations. 3. every 4 generations. 4. once at the
            outset of control; k = 0.10, resistance is a dominant
            trait; snails are released at time 0.

    (4) The model shows that genetic control is most effective
when the initial snail infection rate is high (Figs. 1, 2), because
the selective advantage of resistant snails is then also higher.
However, the model shows that at release ratios of 1:1 and above,
control rates become comparable at the three levels of transmission
considered (Tables 3, 4). This answers criticism that genetic
control will only work if infection by schistosomes is a key factor
regulating snail populations (Michelson and Dubois, 1978), since
sufficiently high release ratios will compensate for low levels of
schistosome transmission. High release ratios may also compensate
for a possible selective disadvantage of resistant snails relative
to susceptible snails (e.g., Minchella and LoVerde, 1983). For
example, the model predicts that if resistant snails have only half
the fitness, on the average, of susceptible snails, a 3:1 ratio
four times a year would still achieve a reduction in the snail
infection rate from 0.1 to 0.000001 within 2 to 4 years. Genetic
control is thus theoretically feasible even if resistance is in
fact associated with deleterious effects.

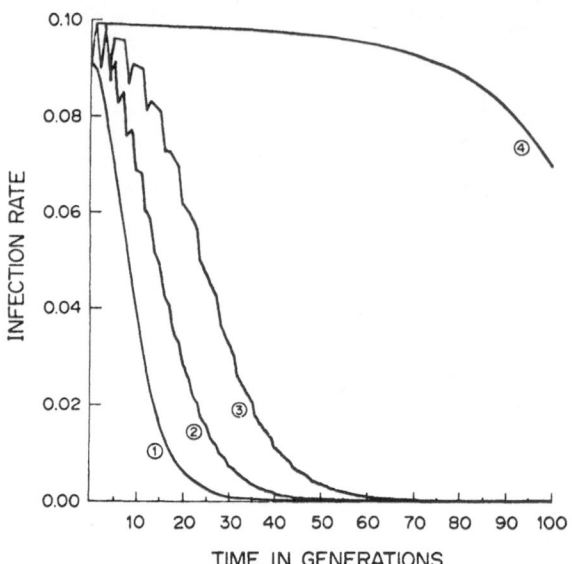

FIGURE 6.   Change in the snail infection rate over time measured
            in snail generations, in response to releases of
            resistant snails at a 1:10 ratio; 1. every generation.
            2. every 2 generations. 3. every 4 generations.
            4. once at the outset of control; k = 0.10, resistance
            is a recessive trait; snails are released at time 0.

        The general conclusions outlined above should hold true even
when many of the assumptions of the model are relaxed. On the
other hand, the model is clearly unsuited in its present form for
making accurate quantitative predictions of what will happen under
different regimes of genetic control. For example, the genetic
basis of schistosome-snail compatibility is certainly more complex
than the monogenic system modeled (Newton, 1953; Richard, 1970,
1975b). The extent to which a multi-locus system would modify
the rate of increase in the proportion of resistant snails would
depend, in part, on the number and degree of dominance of the
genes involved, and on the heritability of resistance, i.e., the
degree to which resistance is controlled by genetic factors. As
shown above, however, the control method relies more on the frequent
release of large numbers of resistant snails than on the operation
of natural selection, so that the rate of change should not be
appreciably decreased if the genetic basis for compatibility is
multigenic rather than monogenic. Seasonal and change fluctuations
in snail and miracidial densities and in environmental conditions,
which affect transmission rates, would have to be taken into

account in a more detailed model. The migration of snails and
schistosomes with different genes for compatibility into the control area would also have to be considered. It may be that genetic
control should be tried first in relatively closed ecosystems such
as static water bodies in isolated rural areas.

Released snails may be less fit than wild snails, either because
resistance is associated with a selective disadvantage, or as a
result of laboratory culture and consequent loss of adaptation to
natural conditions. The problems associated with maintaining a
selective drive favoring the introduced genotypes and of coping
with maintaining a selective drive favoring the introduced genotypes
and of coping with the immigration of new genotypes are well known
to workers involved in plant resistance programs (Day, 1974) and in
the genetic control of insect pests (Pal and LaChance, 1974; Pal and
Whitten, 1974). Several decades of accumulated experience in genetic
control with other groups of organisms should prove to be of help
in devising an effective strategy of genetic control for schistosomiasis. For example, release ratios 10:1 or 50:1 are commonly used in
genetic control of insect pests. Such high release ratios are not
unrealistic in the case of snails if the resident snail population
is first suppressed with molluscicides. The release of resistant
snails can also be timed to coincide with the onset of the breeding
season. Various techniques can be used to improve the field adapt-
ability of resistant strains, including maintenance under natural
conditions.

With improved knowledge of snail genetics, genetic transport
mechanisms such as are used in insect control programs could be
experimented with. Such mechanisms include the introduction of
other favorable genes linked to resistance, meiotic drive (i.e.,
the distortion of segregation ratios with preferential inclusion
in gametes of chromosomes with the desired allele), and cytoplasmic
incompatibility (or sterility of cross-matings between the introduced and native strains).

Prospects for genetic control

The idea of genetic control of schistosomiasis is based on the
premise that the degree of susceptibility of snail populations and
snail infection rates are a significant factor regulating transmission. This hypothesis may meet with skepticism because of the
prevailing view (e.g., Hairston, 1973) that even very low snail
infection rates (0.001-0.01) are sufficient to bring about high
infection rates in the human population. However, the prevalence
of snail infections is often given as an overall average, throughout a year or over many collecting sites. I will argue that such
a way of reporting snail infection rates is misleading, and that
overall averages have little epidemiological significance. This is

because within an endemic area schistosomiasis transmission is not uniform over space and time, but rather is highly focal and seasonal. For example, in any given locality snail infection rates differ markedly from site to site, usually in relation to patterns of human water contact and habitat suitablity for snails (e.g., Gordon et al., 1934; Scott, 1940; Pesigan et al., 1958; Sturrock, 1973; Upatham, 1976; Ukoli and Asumu, 1979). Transmission foci, as determined by cercarial densities, are usually rather limited in size (Rowan, 1965; Theron et al., 1978), and sharp gradients in snail infection rates may occur over short distances in relation to proximity to point sources of contamination (e.g., Faust and Hoffman, 1934; Rowan, 1965; Fine, 1976; Theron et al., 1978). Also, at any given site, important fluctuations in the prevalence of snail infections occur, in response to changes in climatic conditions (reviewed in detail by Jordan and Webbe, 1969; also see Anderson and May, 1979) or due to irregular fecal contamination (Sturrock et al., 1979). Usually snail infection rates reach a peak during the relatively short rainy season, and these peaks may easily be missed unless surveys are made on a regular basis (see Paperna, 1968). Because of such high spatial and temporal heterogeneity in snail infection rates, overall averages can contribute little to an understanding of the relationship between schistosome prevalence in snails and in humans. There is both empirical and theoretical evidence to suggest that in fact, prevalence of snail infections may be more important in determining incidence in humans than has been commonly assumed.

As mentioned above, sites where snail collections are made do not contribute equally to schistosomiasis transmission because of differences in the amount of human water contact and in habitat suitability for snails. Because sites where much human water contact occurs are also likely to be most heavily polluted, snail infection rates are usually also higher (Gordon et al., 1934; Scott, 1940). Based on thoretical considerations, Barbour (1978, 1982) and Anderson and May (1979) have suggested that within a given locality it is precisely the sites where snail infection rates are highest that are overwhelmingly responsible for maintaining disease transmission. Thus, even though the overall prevalence of snail infections in any given area is usually very low, it does not automatically follow that low snail infection rates are sufficient to maintain high incidence in humans. A significant decrease in the snail infection rate may indeed have important repercussions on schistosome transmission to humans. Clearly, the epidemiological importance of snail infection rates and of the degree of susceptibility of snail populations merits more careful attention.

Other problems remain to be solved before genetic control of schistosomiasis can become practical (reviewed by Woodruff, in this volume). In the face of continued expansion of the disease, however, it is important to explore adjuncts to established control methods such as mollusciciding or chemotherapy. Laracuente et al. (1979) point out the prohibitive cost to most tropical countries of control programs based on synthetic chemicals, and suggest that labor-intensive biological and environmental methods may offer economic advantages.

At present, control of schistosomiasis by any one method is unrealistic. On the other hand, a combination of complementary methods would have synergistic effects (MacDonald, 1965; Schad and Rozeboom, 1976; Rosenfield et al., 1977).

The present study suggests that genetic control of schistosomiasis is theoretically feasible. It may prove to be a viable adjunct to other control methods such as chemotherapy, sanitation, health education, and chemical and environmental control of snails, and as such deserves further consideration.

V.   ACKNOWLEDGEMENTS

This paper is based on work carried out at Purdue University under the guidance of D. S. Woodruff and P. T. LoVerde. I thank G. G. Plorin for helpful suggestions on the development of the mathematical model and A. W. Cheever, V. R. Ferris, J. E. Foster, M. E. Gilpin, and F.M.A. Ukoli for comments on the manuscript. Support from a National Science Foundation Graduate Fellowship and from the Edna McConnell Clark Foundation is gratefully acknowledge.

VI. REFERENCES

Anderson, R. M.   (1978).   Population dynamics of snail infection by miracidia. *Parasitology*, 77, 201-224.

Anderson, R. M. and May, R. M.   (1979).   Prevalence of schistosome infections within molluscan populations: observed patterns and theoretical predictions. *Parasitology*, 79, 63-94.

Barbour, A. D.   (1979).   Macdonald's model and the transmission of bilharzia. *Trans. R. Soc. Trop. Med. Hyg.*, 72, 6-15.

Barbour, A. D.   (1982).   Schistosomiasis. *In* "Population Dynamics of Infectious Diseases" (Anderson, R. M., ed.).   Chapman and Hall, London.   180-208 p.

Basch, P. F.   (1975).   An interpretation of snail-trematode infection rates: specificity based on concordance of compatible phenotypes. *Int. J. Parasitol.*, 5, 449-452.

Chernin, E. (1970). Behavioral responses of miracidia of *Schistosoma mansoni* and other trematodes to substance emitted by snails. *J. Parasitol.*, 56, 287-296.

Cohen, J. E. (1977). Mathematical models of schistosomiasis. *Annu. Rev. Ecol. Syst.*, 8, 209-233.

Day, P. R. (1974). Genetics of Host-Parasite Interactions. W. H. Freeman and Co., San Francisco.

Falconer, D. A. (1960). Introduction to Quantitative Genetics. The Ronald Press Company, New York.

Faust, E. C. and Hoffman, W. A. (1934). Studies on *Schistosomiasis mansoni* in Puerto Rico. III. Biological studies. 1. The extra-mammalian phases of the life cycle. *J. Public Health Trop. Med.*, 10, 1-47.

Fine, P.E.M., rapporteur. (1976). Mathematical Models of Schistosomiasis. Proceedings Workshop Bellagio, Italy, 1-14 May 1976. Edna McConnell Clark Foundation, New York. 58 p.

Gordon, R. M., Davey, T. H., and Peaston, H. 1934. The transmission of human bilharziasis in Sierra Leone, with an account of the life cycle of the schistosomes concerned, *Schistosoma mansoni* and *Schistosoma haematobium*. *Ann. Trop. Med. Parasitol.*, 28, 323-418.

Hairston, N. (1973). The dynamics of transmission. *In* "Epidemiology and Control of *Schistosomiasis (Bilharziasis)*(Edited by Ansari, N.), Karger, Basel. 250-336 p.

Jordan, P. and Webbe, G. (1969). Human Schistosomiasis. C. C. Thomas, Springfield.

Laracuente, A., Brown, R. A., and Jobin, W. (1979). Comparison of four species of snails as potential decoys to intercept *Schistosome miracidia*. *Am. J. Trop. Med. Hyg.*, 28, 99-105.

Macdonald, G. (1965). The dynamics of helminth infection, with special reference to schistosomes. *Trans. R. Soc. Trop. Med.*, 59, 489-506.

May, R. M. and Anderson, R. M. (1979). Population biology of infectious diseases: Part II. *Nature*, 280, 455-461.

Meuleman, E. A. (1972). Host-parasite interrelationships between the freshwater pulmonate *Biomphalaria pfeifferi* and the trematode *Schistosoma mansoni*. *Neth. J. Zool.*, 22, 355-427.

Michelson, E. H. and Dubois, L. (1978). Susceptibility of Bahian populations of *Biomphalaria glabrata* to an allopatric strain of *Schistosoma mansoni*. *Am. J. Trop. Med. Hyg.*, 27, 782-786.

Minchella, D. J. and Loverde, P. T. (1983). Laboratory comparison of the relative success of *Biomphalaria glabrata* stocks which are susceptible and insusceptible to infection with *Schistosoma mansoni*. *Parasitology*, 86, 335-344.

Mulvey, M. and Vrijenhoek, R. C. (1981). Multiple paternity in the hermaphoroditic snail *Biomphalaria obstructa*. *J. Hered.*, 72, 308-312.

Mulvey, M. and Vrijenhoek, R. C. (1982). Population structure in *Biomphalaria glabrata*: examination of a hypothesis for the patchy distribution of susceptibility to schistosomes. *Am. J. Trop. Med. Hyg.*, 31, 1195-1200.

Newton, W. L. (1953). The inheritance of susceptibility to infection with *Schistosoma mansoni* in *Australorbis glabratus*. *Exp. Parasitol.*, 2, 242-257.

Pal, R. and Lachance, L. E. (1974). The operational feasibility of genetic methods for control of insect vectors of medical and veterinary importance. *Annu. Rev. Entomol.*, 19, 269-291.

Pal, R. and Whitten, M. (1974). The Use of Genetics in Insect Control. Elsevier/North Holland, Amsterdam.

Paperna, I. (1968). Studies on the transmission of schistosomiasis in Ghana. II. The infection rate of snails at transmission sites. *Ghana Medical J.*, 7, 63-70.

Paraense, W. L. (1955). Self- and cross-fertilization in *Australorbis glabratus*. *Mem. Inst. Oswaldo Cruz*, 53, 285-291.

Persigan, T. P., Farooq, M., Hairston, N. G., Jauregui, J. J., Garcia, E. G., Santos, A. T., Santos, B. C., and Besa, A. A. (1958). Studies on *Schistosoma japonicum* infection in the Philippines. 2. The molluscan host. *Bull. WHO*, 18, 481-578.

Richards, C. S. (1970). Genetics of a molluscan vector of schistosomiasis. *Nature*, 227, 806-810.

Richards, C. S. (1975a). Genetic factors in susceptibility of
    Biomphalaria glabrata for different strains of *Schistosoma
    mansoni*. *Parasitology*, 70, 231-241.

Richards, C. S. (1975b). Genetic studies on variation in
    infectivity of *Schistosoma mansoni*. *J. Parasitol.*, 61, 233-236.

Ritchie, L. S., Berrios-Duran, L. A., and Deweese, R. (1963).
    Biological potentials of *Australorbis glabrata*: growth and
    maturation. *Am. J. Trop. Med. Hyg.*, 12, 264-268.

Rosenfield, P. L., Smith, R. A., and Wolman, M. G. (1977).
    Development and verification of a schistosomiasis transmission
    model. *Am. J. Trop. Med. Hyg.*, 26, 505-516.

Rowan, W. B. (1965). The ecology of schistosome transmission foci.
    Bull. WHO, 33, 63-71.

Schad, G. A. and Rozeboom, L. E. (1976). Integrated control of
    helminths in human populations. *Annu. Rev. Ecol. Syst.*, 7,
    393-420.

Scott, J. A. (1940). Schistosomiasis in irrigated mountain valleys
    of Venezuela. *Am. J. Hyg.*, 21, 1-15.

Slobodkin, L. B. (1961). Growth and Regulation of Animal Popula-
    tions. Holt, Rinehart, and Winston, New York.

Sturrock, B. M. (1961). The influence of infection with
    Schistosoma mansoni on the growth rate and reproduction of
    Biomphalaria pfeifferi. *Ann. Trop. Med. Parasitol.*, 60,
    187-197.

Sturrock, R. F. (1973). Field studies on the transmission of
    Schistosoma mansoni and on the bionomics of its intermediate
    host, *Biomphalaria glabrata*, on St. Lucia, West Indies. *Int.
    J. Parasitol.*, 3, 175-194.

Sturrock, R. F., Karamsadkar, S. J., and Ouma, J. (1979).
    Schistosome infection rates in field snails: *Schistosoma
    mansoni* in *Biomphalaria pfeifferi* from Kenya. *Ann. Trop.
    Med. Parasitol.*, 73, 369-375.

Theron, A., Pointier, J. P., and Combes, C. (1978). An ecological
    approach to the problem of the responsibility of men and
    rats in the workings of a transmission site of *Schistosoma
    mansoni* in Guadeloupe (West Indies). *Ann. Parasitol. Hum.
    Comp.*, 53, 223-234. (In French).

Ukoli, F.M.A. and Asumu, D. I. (1979). Freshwater snails of the
    proposed federal capital territory in Nigeria. *Niger. J. Nat.
    Sci.,* 1, 47-56.

Upatham, E. S. (1976). Field studies on the bionomics of the free-
    living stages of St. Lucian *Schistosoma mansoni. Int. J.
    Parasitol.,* 6, 239-245.

Wagoner, D. E., McDonald, I. C., and Childress, D. (1974). The
    present status of genetic control mechanisms in the house fly,
    Musca domestica L. *In* "The Use of Genetics in Insect Control"
    (Edited by Pal, R. and Whitten, M.), Elsevier/North Holland,
    Amsterdam. 193-197 p.

Webbe, G. (1962). The transmission of *Schistosoma haematobium*
    is an area of Lake Province, Tanganyika. *Bull. WHO,* 27, 59-
    85.

Woodruff, D. S. (1978). Biological control of schistosomiasis by
    genetic manipulation of intermediate host snail populations.
    Proc. Int. Conf. Schistosomiasis, Cairo, Egypt (October 18-25,
    1975), 2, 755.

Woodruff, D. S. (1985). Genetic control of schistosomiasis: a
    technique based on the genetic manipulation of intermediate
    host snail populations. *Comp. Pathobiol.* (This volume).

Wright, C. A. (1971). Comments on the paper "Genetics of a mollus-
    can vector of schistosomiasis" by C. S. Richards. *Trop.
    Dis. Bull.,* 68, 333-335.

REACTION OF BLOOD CELLS IN *Ostrea edulis* AND *Crassostrea gigas*: A

NONSPECIFIC RESPONSE OF DIFFERENTIATED CELLS[1]

G. Balouet and M. Poder

Laboratoire de Pathologie
Faculté de Médecine
29279 Brest Cedex
France

---

[1]This work was supported in part by Grant 82P0044 from the
Ministery of Research and in part by Grant 83-2908 from CNEXO.

I.   INTRODUCTION

During the past 20 or 30 years considerable progress has been
achieved in the field of molluscan hematology, especially in the
identification of the main types of blood cells. In bivalve
molluscs, what is now known can be summarized as follows:

(1) Description in hemolymph of two basic types of hemocytes,
     granular and agranular, respectively, these have been de-
     signated as granulocytes and hyalinocytes. The most common
     equivalent term for the granulocyte is the amoebocyte, and
     that for the hyalinocyte is the hyaline hemocyte or lymphocyte-
     like cell. The latter designation is based solely on
     morphological similarities with vertebrates lymphocytes, without
     functional or immunological implications.

     The third type of hemocyte, i.e., the serous cell or pigment
     cell, is found principally in connective tissue, and is
     apparently different in its histogenesis, morphology, and
     physiology. These cells serve an excretory function.

(2) As a result of the absence of a histologically defined hemato-
     poietic organ (Cheng, 1981) and concrete information relative
     to the stages of cell renewal, several hypotheses have been
     proposed; furthermore, in some bivalve species, descriptions
     of progenitor or stem cells have been contributed (Mix, 1976;
     Moore and Lowe, 1977). Such stem cells have been reported
     scattered or arranged in small groups in connective tissue.
     In our opinion, confirmation, especially by labelling
     techniques, is required before the nature of these cells can
     be accepted. Presently, we can only state that the same cells
     are found in hemolymph vessels, the pericardial cavity, and
     connective tissue, and that no characteristic fixed hemocytes
     occur in bivalves. In these molluscs, exchanges between cir-
     culating and interstitial compartments are favored because of
     the semi-closed (or open) arrangement of the vascular system.

(3) No specific immunologically active substances, such as immuno-
     globulins or complement, have been demonstrated thus far in
     any mollusc (Cheng, 1978b; Bayne et al., 1980) but hemagglu-
     tinins or opsonins apparently occur in some invertebrates.
     Moreover, lysosomal enzyme activities have been demonstrated
     in many molluscan species, especially in bivalves; these
     include esterases, phosphatases, and peroxidases, which are
     clearly present in normal and reactive hemocytes. It has been
     proposed that these enzymes represent one of the most impor-
     tant components of cellular defense (Cheng, 1978a).

As we have stated in a previous paper (Balouet and Poder, 1979), it is very difficult to reconcile the divergent descriptions, function stages, and labels contributed by various investigators and reviewers relative to the hemocytes in different molluscan species (Ratcliffe and Rowley, 1981). In our opinion, at the same time it is necessary to recognize the existence of interspecific morphological and physiological variations, and to emphasize that a very few types of the cellular reactions occur in molluscan inflammatory processes. The involvement of hemocytes in diapedesis (Des Voignes and Sparks, 1968), formation of intravascular clots (Bang, 1973), leucocytosis (Cheng, 1978b), phagocytosis of carmine particles (Mikhailowa and Pradznikov, 1961) and of carbon (Moore and Lowe, 1977), and their participation during exocytosis, along with epithelial excretion, of phagocytosed particles and wound repair (Pauley and Heaton, 1969; Ruddell, 1969; Des Voignes and Sparks, 1968) have been reported. Consequently these cells can be compared with those of the diffuse reaction system (RES) in vertebrates. Furthermore, these cells may be functionally defined as a "system of macrophages".

In the present paper, we are presenting the results of light microscope investigations conducted on oysters either during routine histopathological surveys or under experimental conditions designed to evaluate the importance of granular cell reactions to different injuries.

II. MATERIALS AND METHODS

Physiological and Pathological Investigations

Some 50,000 oysters (about 40,000 European flat oysters, *Ostrea edulis*, and 10,000 Pacific oysters, *Crassostrea gigas*, have been sampled since 1974 from Breton coastal waters, particularly in North Brittany, France. These investigations were perfomed by routine histological and ultrastructural methods during surveys for parasitic diseases or for monitoring oil pollution in a very exposed area, with controls obtained from pollution-free beds. The age of oysters ranged from young spat to 7-years-old specimens.

In an attempt to reproduce experimental inflammatory reactions, we submitted 2-years-old *O. edulis* grown in closed system tanks to (1) injection or implantation of inert materials [talc particles (0.2 ml) suspended in sea water or complete Freund's adjuvant (20%) or spongel (a surgical sterile dry reabsorbing gelatine)]; (2) injections of bacteria [0.2 ml of diluted BCG S (15 mg dry weight)].

In both experiments, the oysters were examined at 6 hr to 60 days.

III. OUR RESULTS COMPARED WITH PUBLISHED DATA

Physiological Reactions

In a "normal" healthy oyster, the number of hemocytes in connective tissue is always very small. However, several authors (Tranter, 1958; Lubet, 1959; Houtteville, 1974) have pointed out the presence of reactive hemocytes around and in the gonads during the cleaning up stage of the sexual cycle, especially in the mussel, *Mytilus edulis*. We have observed the same phenomenon in oysters, especially in *Crassostrea gigas*. Specifically, at the end of the maturation period, gonadal tubes are surrounded and infiltrated by granular hemocytes. Simultaneously, lysis of ovocytes and atrophy of gonadal epithelium occur. Phagocytosis and resorption of mature gonadal cells may occur. These phenomena are always more conspicuous in female than in male molluscs, and they disappear at the termination of the reproductive cycle without any scar formation.

If the cycle is disturbed by external insults, the cellular reaction becomes greater along the North Breton coast. After the *Amoco-Cadiz* oil spill, we found important perturbations of gonadal index and spawning related to destruction of the gonads. This was probably due to the affinity of hydrocarbons for the lipid constituents of germ cells. Granulocytic infiltration was found within tubules but disappeared after complete destruction of ovocytes.

Pathological Insults

After a nonspecific insult inflammatory reaction involving an increased number of hemocytes in the connective tissue is the most common finding in oysters in poor condition. Changes associated with alterations in natural environmental conditions or after transport are usually very difficult to identify. We had the opportunity to study oysters that had been exposed to hydrocarbon pollution. There were large numbers of granulocytes present, expecially around the stomach and intestine. Granular cells, with abundant clear vesicular cytoplasm and small eccentric nuclei usually do not engage in phagocytosis. Hyalinocytes are rarely involved in cellular reactions. Exceptions, however, do occur. For example, the case of the *Amoco-Cadiz* pollution, their number was related to the amount of hydrocarbons in sea water and sediment during the first year after the oil spill.

Parasitic Diseases

Cellular reactions are variable in the case of parasitic
infections, depending on the type of parasites present.

Protozoan infections.   In infections with *Marteilia*, only a
very light reaction occurs. This is the case in flat oysters, *Ostrea
edulis*, infected with *M. refringens* (see Balouet, 1979), in mussels,
Mytilus edulis, *Mytilus* sp., or *Crassostrea commercialis* infected
with *M. sydneyi* (see Wolf, 1979). Cellular reactions only become
pronounced in very advanced cases involving necrosis of the digestive diverticula. Neither phagocytosis nor encapsulation were
observed associated parasites situated in the digestive tubules.

The absence of granulocytic reaction is more surprising in the
case of *Minchinia* infections as it has been reported to occur in
Crassostrea virginica with MSX disease (caused by *Haplosporidium
nelsoni*) by Farley (1968) and found by us in the few cases of
Minchinia armoricana (*Haplosporidium armoricanum*) detected in
Ostrea edulis in Europe. In the case of *Minchinia* infections,
connective tissue appears not to respond to the presence of even
large numbers of parasitic plasmodia, only some hyaline cells
occur around the parasites.

Since the spread during 1979 in European waters of *Bonamia
ostreae* (X parasitosis), many investigators have pointed out the
importance of hemocytic reactions in infected oysters. Such reactions are always present in *O. edulis*, around stomach and vessels,
in mantle, and especially in the gills. In epizootic areas, when
such hemocytic reactions occur in the gills, the disease caused by
B. ostreae must be suspected even if the parasite is not immediately
demonstrable.

We have established in smear preparations as well as in histological sections and by employing electron microscopy that *B.
ostreae* develops in the cytoplasm of granulocytes after being
phagocytosed. It divides in the host cell where up to ten *Bonamia*
may be found. This protozoan is involved in degenerative changes in
the cytoplasm of the host's phagocytic cells. The host cell is
eventually killed, releasing the enclosed parasites (Balouet et al.,
1983). Phagocytosis, intracellular multiplication, followed by
release of the parasite are probably responsible for the dissemina-
tion of *B. ostreae* throughout the entire body of the host via
vessels. The parasites are sometimes found in epithelial cells
lining the stomach where they can cause the development of necrotic
areas or abcesses. Similar observations can be made in both natural
infections and in oysters experimentally infected by exposure in
contaminated holding tanks. Our pathological findings are very
similar to those of Farley (pers. commun.) in *O. edulis* and *C. gigas*

infected by the so-called American "microcell." This is one reason
for establishing comparative studies between European and American
oyster diseases.

Helminth infections. In the case of infections by parasitic
worms, we have observed, especially in cockles parasitized by larval
trematodes, heavy granulocytic infiltration around the parasites.
The granular cells involved frequently exhibit an eosinophilic cytoplasm. Encapsulation and fibroblastic reactions are found primarily
around dead parasites.

Experimental Cellular Reactions

After injection of the talc suspension, we have observed its
transport in peripheral hemolymph vessels and in both gonads and
gills (Poder, 1980; Poder et al., 1982). Early hemocytic reaction
(48 hr) consisted only of granulocytes. Such reactions were extensive and persisted up to 30 days. The talc crystals were present
between granulocytes, without any evidence of phagocytosis. When
the tac was introduced with Freund's adjuvant, severe necrosis
occurred. Similar granulocytic infiltration developed against implanted spongel, which formed a "cell trap" from the 2nd to the 8th
day post-implantation. Connective tissue scar formation ensued,
with superficial reepithelialization after degradation of the
gelatine meshes.

After BCG injection, we notice transport of free bacteria in
hemolymph sinuses, followed by peripheral accumulation. In such
areas, phagocytosis of bacteria by granulocytes was observed as
early as the first day, increased by the 7th day, when the
occurrence of phagocytosed bacteria was confirmed by electron
microscopy. Subsequently, intact bacteria were replaced after 15
days by cytoplasmic lipid droplets. Complete recovery occurred
between 15 and 20 days.

IV. DISCUSSION AND CONCLUSIONS

Inflammation, or inflammatory reaction, constitutes the most
ubiquitous method by which animals express their reaction against
all types of insult. Whatever its philosophical implications may
be (Thomas, 1971), inflammation involves a wide range of elementary
reactions, generally classified as humoral and cellular responses.
Both types of responses can or cannot be directly induced by the
insulting agent. In a challenged host, both types of reactions
can interact, resulting in the production of such substances as
antibodies or complement, or on the contrary, resulting in cell
susceptibility to mediators.

In histopathological observations, the basic difference between morphologically specific and nonspecific inflammatory reactions must be emphasized. In both cases, the succession of elementary stages (i.e., vascular, with congestion and edema; and cellular, with increase of exogenous and endogenous cells followed by detersive activity and regeneration) is the same. But major differences exist in the respective importance of each stage, and existence of characteristic histologic features which allow the diagnosis of "specific inflammation" (Balouet and Leroy, 1972). For example, follicles consisting of lymphocytes, epithelioid cells, and giant cells in homeotherms, and very likely the so-called nodules or granulomas in fish, represent very sophisticated inflammatory organization.

In warm-blooded vertebrates, one can demonstrate a strong relationship between this organization and development in the host of cellular delayed hypersensitivity (Balouet et al., 1975) in lesions as different as those of tuberculosis, parasitosis, fungal or viral infections, or neoplastic stroma. Furthermore, common denominator functions of the cellular reaction is indicated by the occurrence of similar histological features in some skin or lung allergies, and specific foreign body response in patients sensitized, for example, to beryllium (Elias and Epstein, 1962). A comparable situation is found in nodules of fish naturally or experimentally infected with fungi or acid fast bacteria.

Other inflammatory reactions, in which cells are more pleomorphic (i.e., consisting of polymorphonuclear cells, lymphocytes and plasmatocytes, mast cells, and macrophages) and unorganized correspond to nonspecific inflammation, in which humoral factors, and especially antibodies, may or may not intervene.

By comparing the well established observations discussed above, we can point out the main characteristics of natural or experimental inflammations in bivalve molluscs are: (1) The responsiveness of granulocytes is always determined in a very similar way in different species (Cheng, 1978b) and with different insults (Kinne, 1980). Our observations confirm that the differences in number, stage, size, or staining of cytoplasmic granules depend more, as previously mentioned, on the techniques employed and functional stages (Tripp et al., 1966) than on specialization in one or different species (Cheng and Foley, 1975; Cheng, 1981).

Taking into account the conditions of observation (i.e., normal or reaction cells), the description of fresh or fixed cells, the difficulties involved in appreciating the structure and putative functions of the cytoplasmic granules, we are of the opinion that it is better to only emphasize that granulocytes are active, motile, and phagocytic cells, as opposed to hyalinocytes, which demonstrate

little or no phagocytic activity and in which little or no enzymes
have been demonstrated.

It is recalled that Gunthert and Kohler (1964) have demonstrated in the same manner that polymorphonuclear leucocytes in
rabbits and many rodent species include pseudoeosinophilic granules,
often described as being amphophilic in Giemsa-stained materials,
which have the same physiological functions relative to diapedesis,
phagocytosis, as neutrophilic granulocytes in humans.

Based on their morphology and functions, molluscan granular
hemocytes appear to be more akin to vertebrate macrophages (mono-
cytes) than to granulocytes (i.e., polymorphonuclear leucocytes).
Molluscan granular hemocytes may or may not exhibit phagocytic
activity, depending on the nature of the injury or pathogen. This
phagocytosis does not appear to be selective, while in vertebrates
there is a classic distinction between engulfment by cells of the
macrophages-monocyte system which attacks particles measuring more
than 0.5 μm and phagocytosis of smaller particles by granulocytes.

Phagocytosis is in most cases a favorable phenomenon, e.g.,
cleaning up pathogens, necrotic cells, etc. On the other hand,
as in the case of the hemocytic parasitosis caused by *Bonamia
ostreae*, the molluscan phagocytes render possible the distribution
of the parasites throughout the oyster's body via intracytoplasmic
transport through hemolymph vessels. A comparable situation, known
as chronic granulomatous disease, occurs in humans; however, the
human disease is dependent on a disorder in bacterial killing activity (oxidative polymorphonuclear deficiency, detectable by the tetra-
zolium nitro blue test).

We wish to point out that molluscan granular cells are not
(with one exception) involved in hemocytic neoplastic prolifera-
tions which arise from hyaline (or stem ?) cells. In these lesions,
tumor cells have never been found to be phagocytic.

The inflammatory reaction in oysters is always histologically
nonspecific. It can vary overtime, but generally follows the
classic stages as defined in vertebrates. It is noted, however,
that it is difficult to distinguish during the initial phase
between the vascular and cellular stages because of the semi-
closed (or open) topography of the circulatory system in bivalves.

One of the most interesting points in molluscan inflammation
is the persistance of cellular reaction until the complete
elimination of the inflammation-causing agent, especially if it is
a biotic pathogen. The degradation or elimination of foreign bodies
is the first and essential condition leading to scar formation or
tissue restitution. If the foreign body is not eliminated, oyster

tissues engage in encapsulation during which fibroblast-like cells from the capsule, followed by granulocytic infiltration.

Whatever the injury may be, in most cases we have been unable to determine histologically the diagnosis of a disease until the causative agent, e.g., chemical, bacteria, parasites, etc., has been directly demonstrated or isolated from diseased tissues. By definition, these are clearly examples of nonspecific reactions.

No histological differences have been found in specimens naturally or experimentally challenged and in which immunological factors are believed to be involved. Consequently, we are of the opinion that the elementary cellular reaction is very similar for all the parasites we have investigated and we have especially noted that granulocytes react in the same way around alive and dead parasites. This point is probably in disagreement with Kinne's (1980) opinion that: "...living particles tend to induce encapsulation less often and less promptly than do non-living particles". In homeothermic vertebrates, it has been claimed that, in the same way, host tissues do not recognize the presence of live parasites while cellular reactions occur very promptly around dead parasites such as worms, eggs, etc. It may be asked: Is what has been designated as "premunition" or "adaptation" in homeothermic vertebrates present in molluscs?

Relative to our experimental studies, we did not find any specific structure when highly immunogenic materials such as *Mycobacteria* or complete Freund's adjuvant were employed.

In conclusion, molluscan granular cells present in hemolymph vessels and connective tissue, normally quiescent and inactive, can be activated and mobilized in a nonspecific way by insults as different as environmental changes, stress, protozoan or metazoan parasites, bacteria, etc. These molluscan hemocytes can be highly differentiated and at the same time be capable of remarkable reactive adaptativity with proteiform cytological features.

This synthesis of the functions of molluscan blood cells suggests the occurrence of a fundamental difference relative to histopathological lesions from that observed in higher vertebrates, especially the absence of delayed cellular hypersensitivity which is required for the development of specific granulomatous lesions. However, it does not eliminate the possibility that other mechanisms, e.g. enzyme activity, opsonization, and cell to cell interaction, occur.

V. REFERENCES

Balouet, G. (1979). *Marteilia refringens*. Biological cycle ques-
    tions and development of Aber's disease in *Ostrea edulis* L.
    Mar. Fish. Rev., 1, 64-66.

Balouet, G. and LeRoy, J. P. (1972). La réaction giganto cellu-
    laire. *Arch. Anat. Pathol.*, 20, 243-251.

Balouet, G. and Ponder, M. (1979). A proposal for classification
    of normal and neoplastic types of blood cells in molluscs.
    In Advances in Comparative Leukemia Research" (D. S. Yohn
    and B. E. Lapin, eds.), pp. 205-208, Elsevier, Amsterdam.

Balouet, G., Levaditi, J. C., and Relyveld, E. (1975). Le
    granulome immunogéne. *Bull. Inst. Pasteur*, 73, 383-409.

Balouet, G., Poder, M., and Cahour, A. (1983). Hemocytic
    parasitosis. Morphology and pathology of lesions in French
    flat oyster *Ostrea edulis* L. *Aquaculture*, 34, 1-14.

Bang, F. B. (1973). Immune reactions among marine and other
    invertebrates. *Bioscience*, 23, 584-589.

Bayne, C. J., Sminia, T., and Van Der Knaap, W. B. W. (1980).
    Immunological memory: status of molluscan studies. *In*
    "Phylogeny of immunological memory", (J. Manning, ed.).
    pp. 57-640. Elsevier, Amsterdam.

Cheng, T. C. (1978a). ·The role of lysosomal hydrolases in
    molluscan cellular response to immunologic challenge. *Comp.
    Pathobiol.*, 4, 59-71.

Cheng, T. C. (1978b). A study of granuloma formation by molluscan
    cell. *Comp. Pathobiol.*, 4, 97-111.

Cheng, T. C. (1981). Bivalves. *In* "Invertebrate Blood Cells"
    Vol. I. (N. A. Ratcliffe and A. F. Rowley, eds.). pp. 233-300.
    Academic Press, London.

Cheng, T. C. and Foley, D. A. (1975). Hemolymph cells of the
    bivalve mollusc *Mercenaria mercenaria*: an electron microscopi-
    cal study. *J. Invert. Pathol.*, 26, 341-351.

Des Voignes, D. M. and Sparks, A. K. (1968). The process of wound
    healing in the Pacific oyster *Crassostrea gigas*. *J. Invert.
    Pathol.*, 12, 53-65.

Elias, P. M. and Epstein, W. L. (1962). Ultrastructural observations on experimentally induced foreign body and organized
epithelioid cell granulomas in man. *Am. J. Pathol.*, 52,
1207-1216.

Farley, C. A. (1968). *Minchinia nelsoni* (Haplosporida,
Haplosporidiidae) disease syndrome in the American oyster
Crassostrea virginica. *J. Protozool.*, 15, 585-599.

Güthert, H. and Köhler, U. (1964). Recherches sur l'origine des
cellules leucocytaires dans l'inflammation. *Ann. Anat. Pathol.*,
9, 49-55.

Houtteville, P. (1974). Contribution à l'étude cytologique et
expérimentale du cycle annuel du tissu de réserve du manteau
de *Mytilus edulis* L. Thése de 3é cycle, Biologie Animale
Caen, France.

Lubet, P. (1959). Recherches sur le cycle sexuel et l'émission
des gamétes chez les Pectinidés et les Mytilidés. *Rev.
Trav. ISTPM* (Paris), 23, 396-545.

Mikhailova, I. G. and Pradznikov, E. V. (1961). Two questions
on the morphological reactivity of mantle tissues in *Mytilus
edulis* L. *T. R. - Murmansk. Morsk. Biol. Inst.*, 3, 125-130.

Mix, M. C. (1976). A general model for leucocyte renewel in bivalve molluss. *Mar. Fish. Rev.*, 38, 37-41.

Moore, M. N. and Lowe, D. M. 1977. The cytology and cytochemistry
of the hemocytes of *Mytilus edulis* and their responses to
experimentally injected carbon particles. *J. Invert. Pathol.*,
29, 18-30.

Kinne, O. (ed.). (1980). "Diseases in Marine Animals." Vol. 1,
J. Wiley and Sons, New York.

Pauley, G. B. and Heaton, L. H. (1969). Experimental wound repair
in the fresh water mussel *Annodonta oregonensis*. *J. Invert.
Pathol.*, 13, 241-249.

Poder, M. (1980). Les réactions hémocytaires inflammatoires et
tumorales chez *Ostrea edulis* L. (Essai de classification des
hémocytes des mollusques bivalves). Thése de 3é cycle,
Océanographie Biologique, Université de Bretagne Ocidentale
Brest, France.

Poder, M., Cahour, A., and Balouet, G.   (1982).   Réactions
    hémocytaires à l'injection de corps bactériens ou de sub-
    stances inertes chez *Ostrea edulis* L.   *Malacologia*, 22, 9-14.

Ratcliffe, N. A. and Rowley, A. F. (eds.).   (1981).   "Invertebrate
    Blood Cells."   Vol. 1, Academic Press, London.

Ruddell, C. L.   (1969).   A cytological and histochemical study of
    would repair in the Pacific oyster *Crassostrea gigas*.   Ph.D.
    thesis, University of Washington, Seattle.

Thomas, L.   (1971).   Adaptative aspects of inflammation.   *In*
    "Immunopathology of Inflammation" (B. K. Forscher and J. C.
    Houck, eds.).   pp. 1-10.   Excerpta Medica, Amsterdam, Holland.

Tranter, D. J.   (1958).   Reproduction in Australian pearl oysters.
    II. *Pinctada albina* (Lamarck).   *Austral. J. Mar. Freshwat.*
    Res., 9, 144-158.

Tripp, M. R., Bisignani, L. A., and Kenny, M. J.   (1966).   Oysters
    amoebocytes in vitro.   *J. Invert. Pathol.*, 8, 137-140.

Wolf, P. H.   (1979).   Life cycle and ecology of *Marteilia sydneyi*
    in the Australian osyter *Crassostrea commercialis*.   *Mar. Fish.*
    Rev., 41, 70-72.

PHAGOCYTOSIS OF INERT PARTICLES: A COMPARATIVE STUDY IN INSECTS

AND MARINE CRUSTACEANS

M. Brehelin

Laboratoire de Pathologie Comparée
Equipe de Recherches en Pathologie des Animaux Marins
U.S.T.L.
Pl. E. Bataillon 34060 MONTPELLIER Cedex
FRANCE

J. M. Arcier

Laboratoire de Physiologie des Invertébres
Equipe de Recherches en Pathologie des Animaux Marins
U.S.T.L.
Pl. Bataillon 34060 MONTPELLIER Cedex
FRANCE

I. INTRODUCTION

As in mammals or other vertebrates, invertebrates are able to recognize and to segregate or eliminate foreign bodies that are introduced into the hemocoel or into tissues. Relative to micro-organisms such as bacteria or viruses, some are recognized as foreign bodies and are discarded while others can develop in the invertebrate's tissues: the latter are usually pathogens or commensals. How insects or other arthropods recognize and engulf foreign particles present in their bodies is poorly understood. Although some studies have shown that molecules such as agglutinins exist in the blood of insects (see Lackie, 1981 for review), the role of these substances *in vivo* in recognition and engulfment by phagocytes is uncertain. It is not clear if free macrophages are attracted by the foreign bodies or if the initial contact is established at random.

As an initial approach to addressing some of the problems relative to phagocytosis, we have compared the phagocytic cells of crustaceans to those of insects, especially their structural features, distribution in the body, and the first events of the engulfment process. We think that a comparative study is better than one focused on a single species. For example, it permits us to state if a distribution pattern or an observed reactions is an isolated phenomenon or if it corresponds to a trend that has been maintained in the course of evolution, i.e., it represents an adaptive progressive trend for the animals.

In the present study, we are comparing recent findings relative to crustacean species to data pertaining to insects, which have, in part, been published earlier (Brehélin and Hoffmann, 1980). In order to avoid complicating the results by such factors as metabolites produced by living materials or by surface molecules which differ from one microorganism to another (see Brook and Krier, 1978), we injected inert (nonimmunogen) particles of different sizes: iron saccharate and latex beads. Iron saccharate is easily recognized by light microscopy with Perl's reaction for ferric iron. With electron microscopy, it exhibits a natural opacity to electrons which permits its detection easily.

II. MATERIALS AND METHODS

The insects employed were fifth instar larvae of *Locusta migratoria* (Orthoptera) and last instar larvae of *Galleria mellonella* (Lepidoptera). *L. migratoria* were reared in the gregaria phase at a relative humidity of 60% and a photoperiod of 12 hr light and 12 hr dark. *G. mellonella* were reared at 28° C on pollen and wax. The two decapods crustaceans were adults of *Palaemon adspersus* (Palaemonidae) and post-larvae of *Penaeus japonicus* (Penaeidae). *P. adspersus* were collected from the Thau pond, Hérault, France. We employed animals of a same size (5 cm). *P. japonicus* was provided by the rearing farm, DEVA-Sud (CNEXO-Palavas), which we want to thank.

In our experiments, iron saccharate or latex beads were suspended in a 1% sterile NaCl solution for injection in insects. Each animal received 10 µl of this suspension consisting of 0.25 mg of inert powder or 5.10^6 latex beads (0.8 to 1 µm diameter). In the case of the crustaceans, iron saccharate or latex beads were suspended in a 3% sterile NaCl solution and 100 µl were injected, i.e., 2.5 mg of inert powder or 5.10^7 beads. Injections were performed in the abdominal hemocoel of insects and in the pericardium of crustaceans.

For electron microscopy we use the following fixatives: (1) For insects, 5% glutaraldehyde solution buffered at pH 7.4 with 0.1 M phosphate buffer and 1% osmium tetroxide in the same buffer. (2) For crustaceans, 3% glutaraldehyde buffered at pH 7.4 with a 0.1M cacodylate and 0.25 M NaCl and 1% osmium tetroxide in the same buffer.

After dehydration in an ethanol series and propylene oxide, the tissues were embedded in Arldite (insects) or Epon 812 (crustaceans). Semithin sections were stained with hematoxylin and eosin, or with the Perl's reaction for ferric iron (Martoja and Martoja, 1967). Ultrathin sections were treated with uranyl acetate followed by lead citrate and examined in a Siemens Elmiskop or a JEOL 100B electron microscope operated at 60 kV and 80 kV, respectively.

III. RESULTS

If we compare phagocytosis of inert particles in the two
arthropod classes studied, similarities and differences can be
appreciated relative to the morphology and disposition of the
phagocytic cells as well as the process of phagocytosis.

The Phagocytes

In a first experiment, 1 hr and 1 day after injection, we removed and fixed different tissues of the animals to see where the
injected particles were localized.

Similarities between Insects and Crustaceans

Types of phagocytic cells. In *L. migratoria* the injected
inert particles are taken up by three different types of cells:
(1) circulating hemocytes; (2) reticular cells of the hematopoietic
tissue (Hoffmann et al., 1968a), described as the phagocytic organ
by Cuénot (1896), and (3) pericardial cells (= nephrocyte-like cells).
For details concerning the third type of cell, see the Crossley
(1972, 1983). Similarly, three different types of cells engulfed
the foreign particles injected in *P. adspersus*: (1) blood cells;
(2) fixed phagocytes; for instance, those of the digestive gland
(Cuénot, 1905); and (3) podocytes in the gills and antennal gland).

Disposition of Phagocytic Cells. In *L. migratoria* as well as
in *P. adspersus*, the phagocytic cells were not disposed at random
among the internal organs. Rather, they are localized in parts of
the body where they were most apt to come in contact with foreign
particles in the hemolymph, i.e., macrophages and pericardial cells
in insects that line the dorsal blood vessel (=heart) near its
ostia where hemolymph enters the heart (Fig. 1), and fixed phagocytes of crustaceans that line the arterioles of the digestive
gland (Fig. 2).

Ultrastructure of Nephrocyte-like Cells. In insects nephrocyte-like cells were localized along the dorsal blood vessel while
in crustaceans they were distributed especially in gills and in
the antennal gland (Figs. 3,4). In both insects and crustaceans
the nephrocyte-like cells exhibited similar ultrastructural
features. In particular, their plasma membrane showed numerous deep
infoldings which formed channels. Each channel appeared to be
closed by a diaphragm laying between two foot-like processes
at the periphery of the cell (Figs. 5,6) as in the podocytes of the
mammalian kidney.

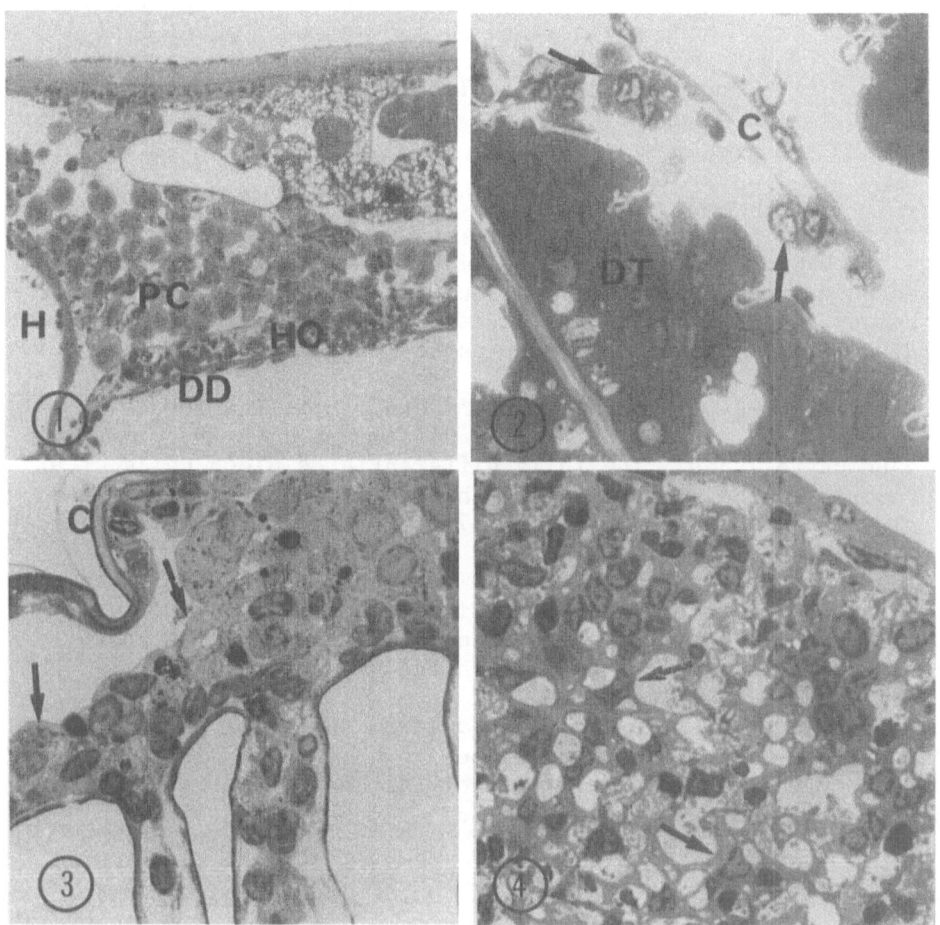

FIGURE 1.   Pericardial sinus of *Locusta migratoria* (Insecta).
            The hemopoietic organ (HO) in which are localized the
            fixed phagocytes is disposed along the dorsal diaphragm
            (DD) near the heart (H). The pericardial cells (PC)
            are regrouped around the dorsal blood vessel (Semithin
            section; X400).

FIGURE 2.   Interstitial tissues in digestive gland of *Palaemon
            adspersus*. The fixed phagocytes (arrows) line the
            capillary (C) between the digestive tubules of the
            hepatopancreas (DT).   Semithin section; X800).

                                      (Continued overleaf)

FIGURES 3 and 4.   Nephrocyte-like cells (arrows) (=podocytes) in
                   the gills (Fig. 3) and antennal gland (Fig. 4)
                   of *P. adspersus*. The lacunae between podocytes
                   are filled with iron saccharate (2 hr post-
                   injection). C = cuticle. (Semithin sections;
                   X800).

Kinds of Engulfed Particles.   In insects as in crustaceans,
iron saccharate was engulfed by the three cells types (hemocytes,
fixed phagocytes, and podocytes) but latex beads were only observed
in hemocytes and in fixed phagocytes; they were absent in the pericardial cells of insects in the "podocytes" of crustaceans.

Differences Between Insects and Crustacean Phagocytes

Hemocyte Types.   The fully differentiated blood cells of *L.
migratoria* can be separated into three distinct types (Hoffmann
et al., 1968b; Brehélin et al., 1978). These are described below.

Macrophagous plasmatocytes.   These are heavily digitated cells
with numerous lysosomes and other inclusions of a resorptive
nature. They appear to be macrophages and their ultrastructure is
close to that of reticular cells engaged in endocytosis.

Granulocytes.   These contain a large number of uniformly dense
granules and have a poorly developed rough endoplasmic reticulum.
Few endocytotic vesicles are evident, although they were not
observed to have endocytosed the injected materials.

Coagulocytes.   These contain both uniformly dense granules
and multitubular globules and have a well developed rough endoplasmic reticulum. Their plasma membrane show numerous pinocytotic
vesicles and pronounced digitations. In this insect species, some
coagulocytes undergo considerable transformation *in vitro*. These
cells are very active phagocytes.

G. mellonella larvae do not possess granulocytes, but in
addition to the hemocytes present in *Locusta*, they have typical
plasmatocytes, oenocytoids, and spherule cells. These last three
types of hemocytes are not phagocytes.

In the two crustaceans species studied, the originality of
each hemocyte type is less evident than in insects. We were able
to distinguish: (1) Hemocytes without dense granules [=hyaline
hemocytes of Bauchau (1981) as reported in other crustaceans
species] which are rare in the circulating hemolymph. Their ultrastructure is different of that of insect plasmatocytes because

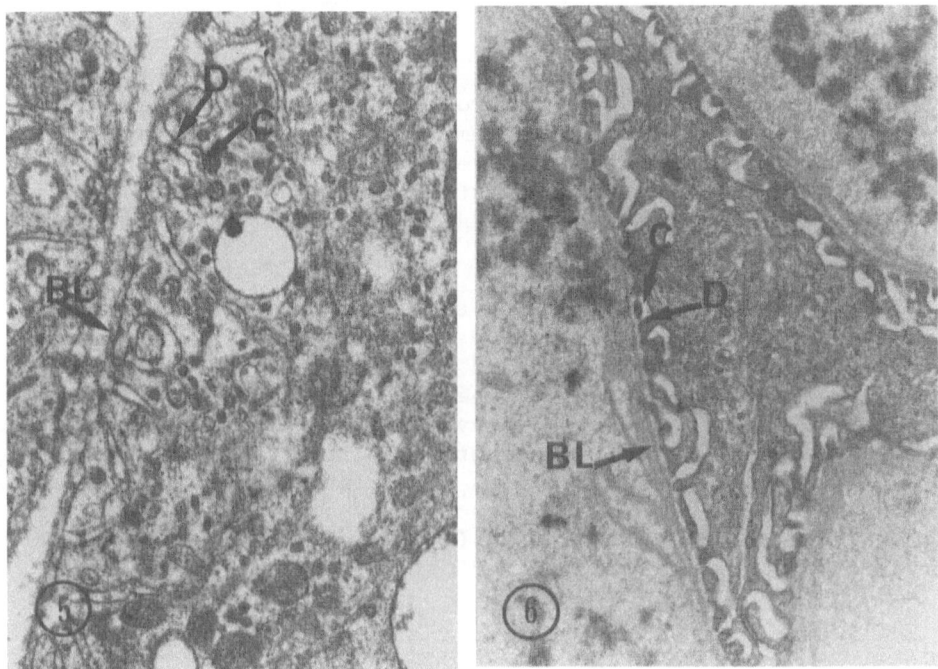

FIGURES 5 and 6.   "Podocytes" (=pericardial cells) of insect (Fig.
                    5) and crustacean (Fig. 6). The cell periphery
                    exhibits numerous foot processes; the deep un-
                    foldings of the plasma membrane result in
                    channels (C) which are closed by the pore dia-
                    phragms (D). These cells are surrounded by a
                    basal lamina (BL). 5. *L. migratoria* X12,000
                    (TEM). 6. *P. adspersus* X12,000 (TEM).

they lack pinocytotic vesicles, pseudopods, and lysosomes. Their
endocytotic capabilities are low, if any.

    (2) Hemocytes with small and rounded granules [=semigranular
hemocytes (?)] which are the most frequently encountered type of
cells of the circulating hemolymph. They seem to be the most fragile
type of crustacean hemocytes *in vitro*. They differ from coagulocytes of insects in that they have a less developed rough endoplasmic reticulum and they do not include structured globules.
Furthermore, their pinocytotic vesicles and inclusions of resorbed
materials are, under normal conditions, much less numerous than

those of insect coagulocytes. In crustaceans, these cells are
the most actively phagocytic.

   (3) Hemocytes with large numerous granules [=granular
hemocytes]. These cells are less numerous than the semigranular
hemocytes. Hemocytes of this type resemble granulocytes of insects.
Like insect granulocytes, these crustacean cells participate in
capsule formation around large foreign bodies. They portray reduced
endocytotic capacities.

   Ultrastructure of fixed phagocytes. In *L. migratoria*, fixed
phagocytes are situated close to the reticular cells of the hematopoietic organ. They are distributed among differentiating hemocytes
and other cells in this tissue (Bréhélin and Hoffmann, 1980;
Zachary et al., 1981) (Fig. 7). They portray similarities with the
histiocytes of vertebrates. In crustaceans, they are arranged in
groups of few cells along arterioles of the digestive gland (Johnson,
1980). Each group of cells is surrounded by a well developed basal
lamina which exhibits numerous gaps (Fig. 8). This layer of basal
material does not exist around fixed phagocytes of insects.

Engulfment Process of Inert Particles

   The engulfment by hemocytes of insects and crustaceans is
different in many ways not reflected by their morphology. In
this study we fixed tissues from 5 min to 7 days after injection
of the foreign particles.

Similarities in Insects and Crustaceans

   In both insects and crustaceans, most of the engulfed particles
occur in fixed macrophages, i.e., reticular cells of the hematopoietic organ in *L. migratoria* (Fig. 9) and fixed phagocytes which
line the arterioles of the digestive gland in *P. adspersus*
(Fig. 10).

   In both insects and crustaceans the process of recognition of
foreign particles and their binding to phagocytes appear to be
the same. Very early after the injection of latex beads, they
become embedded in a floculent material, probably proteinaceous in
nature. This phenomenon is observed in the insects *L. migratoria*
and *G. mellonella* (Fig. 11) as well as in the crustaceans *P.
japonicus* (Fig. 12). It is this material which binds to the
plasma membrane of the phagocytes prior to internalization.

Differences in Engulfment Process

   In insects, engulfment of small inert particles, e.g., iron
saccharate, is accomplished by pinocytotic vesicles which are very

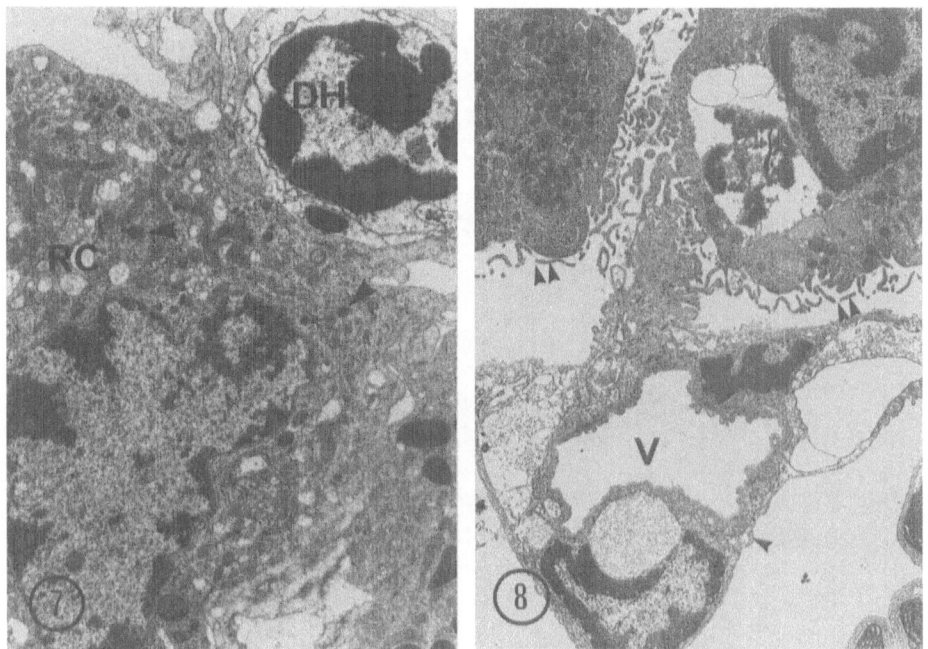

FIGURES 7 and 8.   Disposition of the fixed macrophages in normal
                    animals.   7.   Hemopoietic organ of *L. migratoria*.
                    The reticular cell (RC) has phagocytosed a
                    degenerative hemocyte (DH).   Note the numerous
                    lysosomes present in the cytoplasm of the
                    reticular cell.   Note presence of a blood cell
                    (granulocyte) at lower right.   (TEM; X8,000).
                    8.   Interstitial tissues of hapatopancreas of *P.
                    adspersus*.   Two fixed phagocytes are associated
                    with a blood vessel (V).   Note that the basal
                    lamina of the endothelium is continuous (single
                    arrows) whereas that of phagocytes is interrupted
                    (double arrows) (TEM; X5,000).

numerous at the periphery of all types of phagocytic cells (Fig.
13).   In crustaceans, the bulk of engulfed iron saccharate is
endocytosed in large aggregates which are surrounded by "digitations"
of the cells (i.e., phagocytosis *sensu stricto*) (Fig. 14).

     In insects as well as in crustaceans, latex beads are engulfed
by phagocytosis.   The plasma membrane facing each bead invaginates

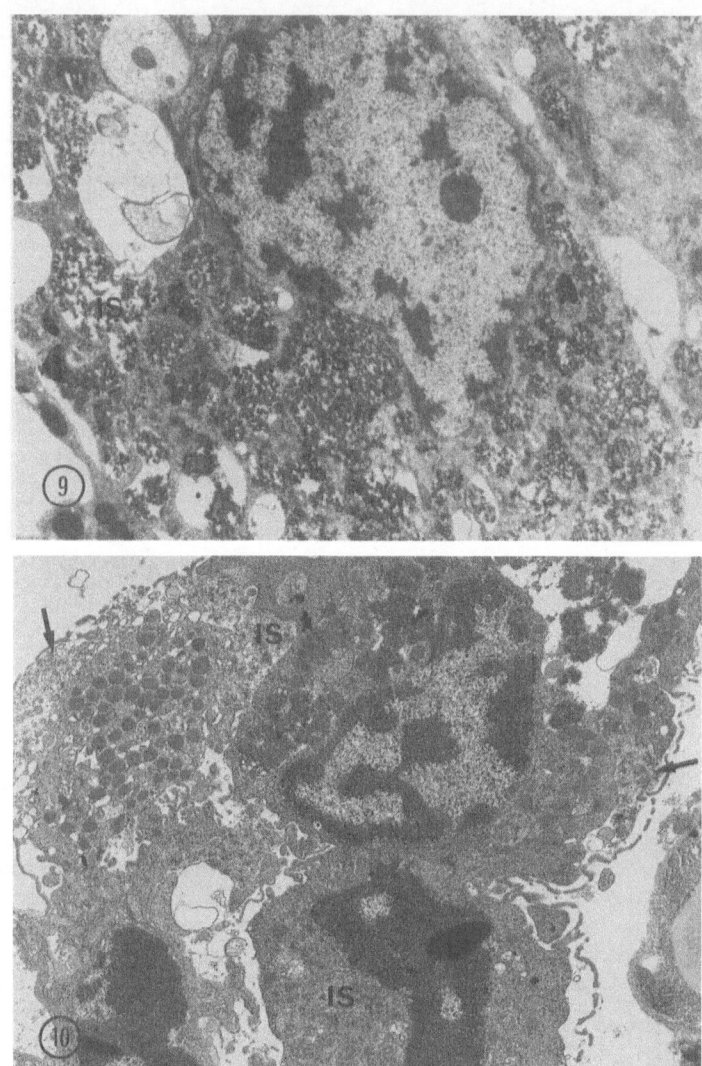

FIGURES 9 and 10.   Fixed phagocytes after injection of iron
                     saccharate (IS).   9.   Reticular cell of the
                     hemopoietic organ of *L. migratoria*.   This cell is
                     overloaded with the inert powder.   (TEM; X10,000).
                     10.   Fixed phagocyte of the hepatopancreas of
                     P. adspersus.   The amount of engulfed inert powder
                     is much less than in cells of *L. migratoria*.
                     Note that large quantities of iron saccharate
                     are concentrated between the cells and the
                     interrupted basal lamina (arrow) (TEM; X8,000).

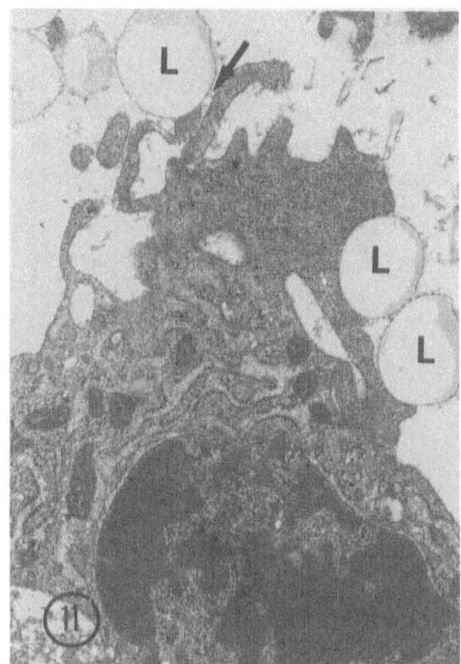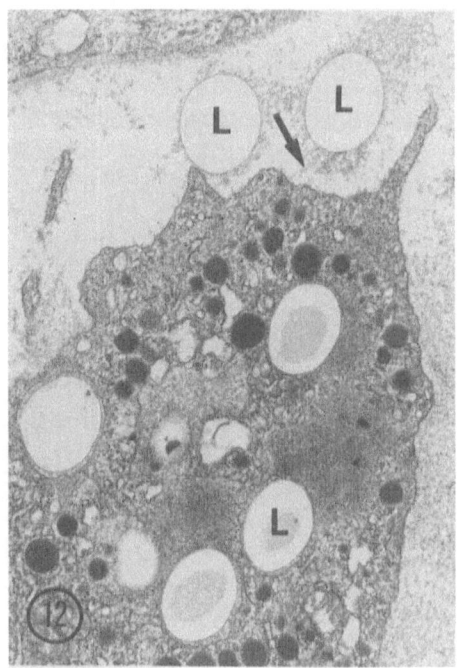

FIGURES 11 and 12.   Recognition of latex beads by circulating
                     hemocytes.   Each bead (L) is coated with a
                     floculent material which binds to the plasma
                     membrane of hemocytes (arrows).   11.   *L.
                     migratoria*, pellet of centrifuged circulating
                     hemocytes, 2 min post-injection. (TEM; X11,000).
                     12.   Gills of *Penaeus japonicus* 5 min post-
                     injection.   (TEM; X9,000).

and the particle seems to enter the invagination.   In insects, one
or several pinocytotic vesicles occur at the bottom of this invagination (Fig. 15).   Such vesicles were not observed in crustaceans
(Fig. 16).

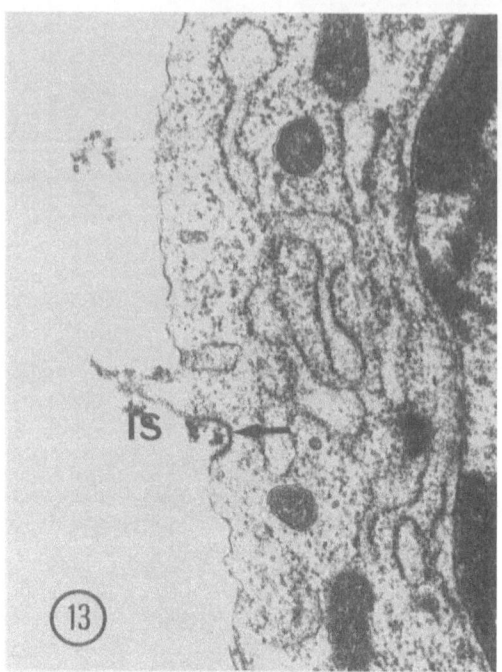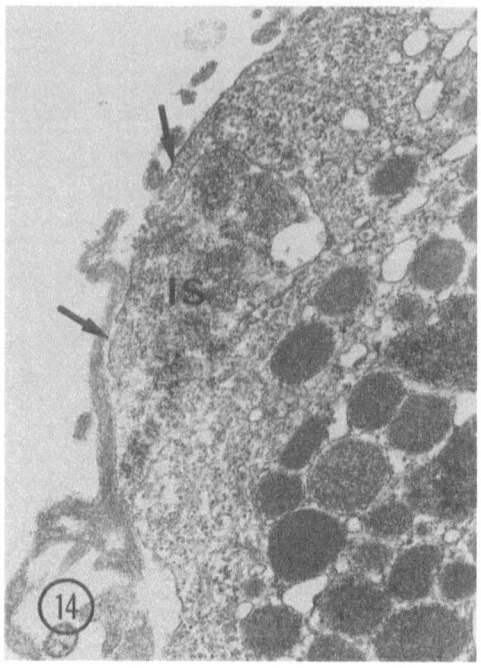

FIGURES 13 and 14.   Mode of engulfment of injected iron saccharate
                      (IS).   13.   Phagocyte of *L. migratoria*; small
                      aggregates of inert powder are taken up by
                      pinocytotic vesicles (arrows). (TEM; X18,000).
                      14.   Macrophage of the interstitial tissues in
                      hepatopancreas of *P. adspersus*; a large amount
                      of iron saccharate is engulfed by a phagocytosis
                      process.   Arrow=digitation of the plasma membrane
                      which surrounds the powder. (TEM; X16,000)

The most important difference between insects and crustaceans
rests with the fate of the injected particles.

In insects, 2 hr after injection, all of the particles (iron
saccharate and latex beads) occurred in vacuoles of the different
types of phagocytes, but especially hemocytes and reticular cells
of the hematopoietic organ.   Moreover, in insect nephrocytes, the
iron saccharate was accumulated in cytoplasmic vacuoles but some
of the iron saccharate was associated, as soon as 1 day post-
injection, with the rough endoplasmic reticulum where it would be

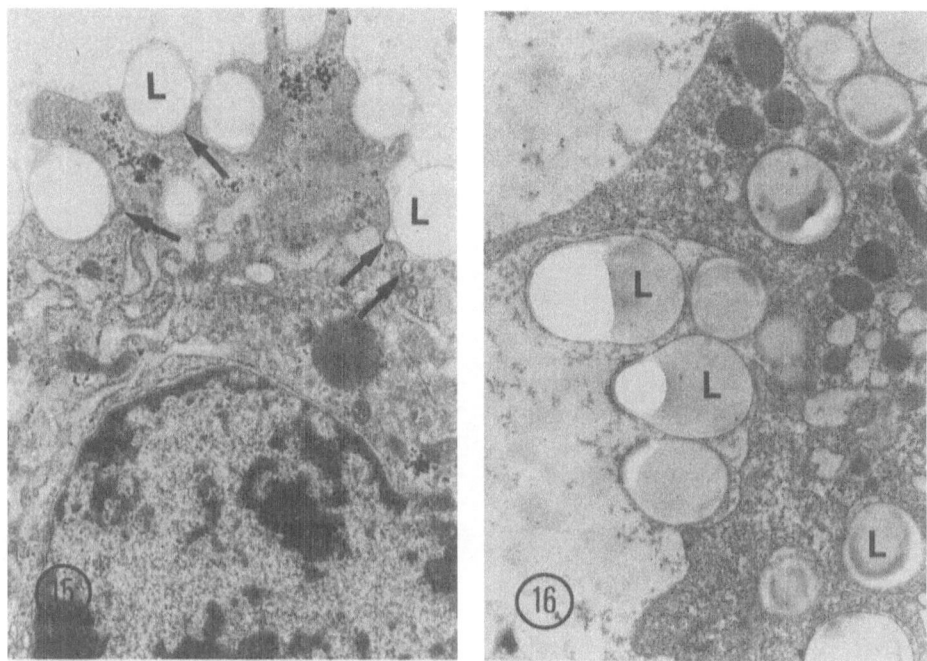

FIGURES 15 and 16.   Phagocytosis of latex beads (L) by circulating
                     hemocytes. 15. Macrophagous plasmatocyte of
                     L. migratoria: Note numerous pinocytotic
                     vesicles (arrows) in bottom of depression which
                     is the beginning of formation of the phagocytic
                     vesicle. (TEM; X11,000). 16. These pino-
                     cytotic vesicles are not observed in *P. adspersus*
                     (semi-granular hemocyte in the gills). (TEM;
                     X11,000).

linked to proteins (Figs. 17, 18). This phenomenon was not observed in crustaceans.

    In crustaceans even a few days after injection, the bulk of the
inert particles agglomerated extracellularly in hemal spaces of the
gills and antennal gland.

    For each type of phagocytic cells, the volume of material
engulfed is considerably greater in insects than in crustaceans
(compare Figs. 9 and 10).

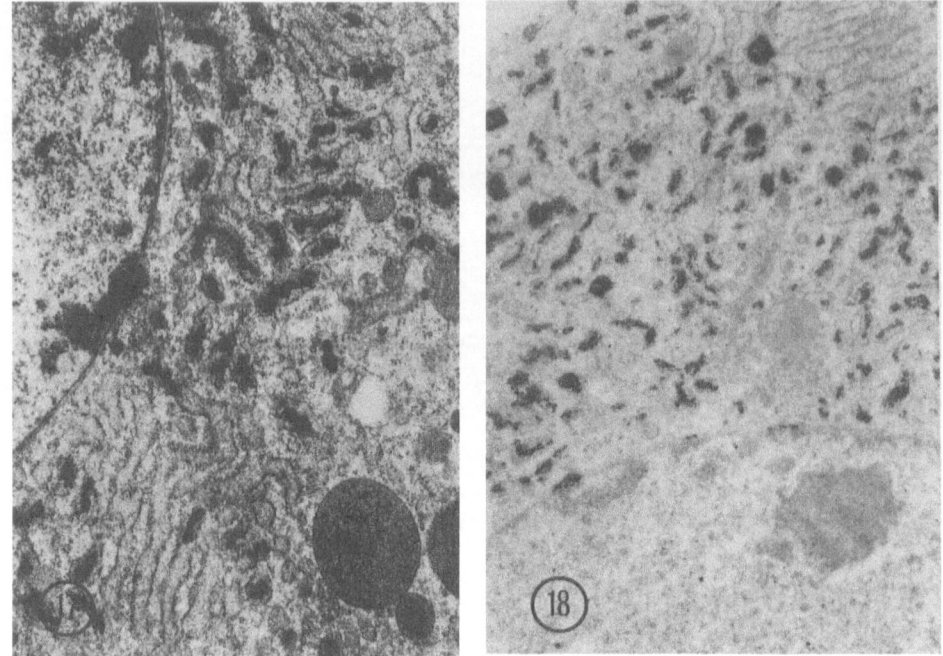

FIGURES 17 and 18.   Pericardial cells of *L. migratoria*, 5 days
                      after injection of iron saccharate. The inert
                      powder is recovered in the vesicles of the RER;
                      the natural opacity of iron saccharate to
                      electrons is evidenced on the uncontrasted
                      section in Fig. 18.   (TEM; X18,000).

     Concerning hemocytes, we wish to emphasize that while all of
the phagocytic hemocytes in insects are loaded with engulfed
particles, a very low percentage of circulating phagocytes of
crustaceans include engulfed material.

IV. DISCUSSION

     In phagocytosis of injected foreign bodies, it is certain
that some of the differences between insects and crustaceans are
due to the anatomy of their circulating systems. Insects have an
hemocoel and only one vessel (dorsal vessel or heart); crustaceans
as *P. adspersus* and *P. japonicus* have, together with hemal lacunae,
a well organized circulating system with a heart, arteries, veins,

and capillaries. Therefore, the distribution of phagocytic cells in the bodies of insects is different from that in crustaceans. Pericardial cells and fixed macrophages of insects are scattered all around their single vessel while the nephrocytes of crustaceans line the long and narrow lacunae of gills and antennal glands. The macrophages of crustaceans are scattered along capillaries of the digestive gland or associated with the heart in some species of shrimps (Cuénot, 1905).

Our experiments show that soon after injection into crustaceans, the bulk of injected iron saccharate or latex beads is found outside cells, in the lacunae of gills and antennal glands where they appear to be entrapped in a proteinaceous matrix. In insects we have never observed such durable concentrations of extracellular foreign material in any part of the body. This difference can be related to differences in anatomy. The other main difference between insects and crustaceans, however, cannot be attributed to differences in the anatomy of their circulatory systems. It reflects the effectiveness of phagocytosis by all of the phagocytes, especially the circulating hemocytes. The weak engulfing capability of crustacean hemocytes have already been emphasized by several authors who think that it would be compensated by the large number of circulating cells (Smith and Ratcliffe, 1978). We have demonstrated that even 7 days after injection of inert particles into *P. adspersus* or *P. japonicus*, the foreign particles can be seen in the hemolymph outside the cells. This is in agreement with Fontaine and Lightner (1974) who studied *Penaeus setiferus*. Moreover, many of our micrographs show circulating hemocytes of the three types in crustaceans in close contact with inert material but not engulfing it. On the other hand, in the two species of insects we have studied, a large number of hemocytes are able to engulf foreign particles immediately after injection and 2 hr later, very few, if any, foreign particles occur free in the hemolymph. That means that a large number of insect blood cells are active phagocytes whereas most of the crustacean hemocytes are not phagocytes or are at least nonactive ones. Moreover, the ability to phagocytose of the nephrocyte-like cells and fixed phagocytes of crustaceans is considerably less than those of insects. This low efficiency, together with anatomical differences, explain why large amounts of injected material are always observed in crustacean lacunae.

We wish to emphasize that the fenestrated basal laminae around fixed phagocytes of the digestive gland in *P. adspersus* do not permit the passage of large particles such as latex beads from the hemocoel towards the macrophages. It is probable that the beads we observed in these cells can only have come from the blood in capillaries. Moreover, concentrations of iron saccharate are always observed between the fenestrated basal lamina and the plasma membrane of macrophages (Fig. 10) which suggests that it does

not come from the hemocoel but from the lumen of the vessel. This
would confirm the work of Cuénot (1905) who believed that hepatic
vessels of crustaceans can open directly in the nodules of phagocytes. This hypothesis has not received support since the publication of Cuénot's work (Johnson, 1980).

Two major similarities exist between insects and crustaceans
relative to phagocytosis. The first is the presence of fixed phagocytes of two types: macrophages and nephrocyte-like cells. In
these groups of arthropods the fixed phagocytes exhibit some
similar ultrastructural characteristics especially the pericardial
cells of insects and the podocytes of crustaceans both of which
possess similar pedal processes with pore diaphragms. Also, the
nephrocyte-like cells engulf particles of small size (in our experiments iron saccharate but not latex beads). The designation of
macrophage some times employed for these cells in the literature
is not adequate.

The second similarity concerns the process of recognizing inert
particles and their binding to the macrophage plasma membrane which
seem to be the same in insects and crustaceans. When living (or
immunogen) foreign material is engulfed by macrophages one can
invoke recognition of some membrane component of this foreign
material by macrophage receptors. It cannot be the same with inert
(nonimmunogen) foreign material. It is certain that the shape,
size of the particles, and their surface properties (electric
charge, etc) are of primary importance (Vinson, 1974; Lackie, 1983).
But in our *in vivo* experiments it appears that the recognition process occurs in two steps. The first is the precipitation and
coating of plasma proteins(?) on the wall of the particles; in this
step no doubt that surface properties interfere. The second step
is recognition of this coat by macrophages and linkage to the plasma
membrane. Here it is not certain that the particle surface properties play an important role because they are probably masked by the
recently deposited coating. Subsequently, the process of ingestion
commences.

What is the origin of this coating material? Two hypotheses
may be advanced. The first is that plasma proteins spontaneously
precipitate upon making contact with inert particles. The second
is that some hemocytes lyse after making contact with foreign
material and induce the formation of the precipitate which coats
the particles. Such a process has been observed in insects
(Rowley and Ratcliffe, 1976) and perhaps in crustaceans (Smith and
Ratcliffe, 1980). But in each case, the precipitate not only coats
the particles but also entraps and segregates them in a proteinaceous
matrix. In *L. migratoria*, in the process of encapsulation of
large foreign bodies, we have observed as a first step the transformation of coagulocytes in contact with recently implanted foreign

bodies (Brehélin et al., 1975; Brehélin, 1977). This transformation leads to the precipitation of some plasma proteins which can adhere on the surface of the foreign material. As a working hypothesis, we propose that the same process occurs prior to phagocytosis of foreign materials sufficiently small to be endocytosed.

V. ACKNOWLEDGEMENTS

We wish to thank Professor J. Hoffmann for his input in the design of some of the experiments and for reviewing the manuscript.

This work has been supported by a grant from CNEXO No. 82/2698.

VI. REFERENCES

Bauchau, A. G. (1981). Crustaceans. In "Invertebrate Blood Cells" (N. A. Ratcliffe and A. F. Rowley, eds.), pp. 385-420. Academic Press, London.

Brehélin, M. (1977). Etude morphologique et fonctionnelle des hémocytes d'Insectes. Thèse d'Etat, Strasbourg. No. 1065.

Brehélin, M., Zachary, D., Hoffmann, J. A., Matz, G., and Porte, A. (1975). Encapsulation of implanted foreign bodies by hemocytes in Locusta migratoria and Melolontha melolontha. Cell. Tiss. Res., 160, 283-289.

Brehélin, M., Zachary, D., and Hoffmann, J. A. (1978). A comparative ultrastructural study of blood cells from nine insect species. Cell. Tiss. Res., 195, 45-57.

Brehélin, M. and Hoffmann, J. A. (1980). Phagocytosis of inert particles in Locusta migratoria and Galleria mellonella: study of ultrastructure and clearance. J. Insect Physiol., 26, 103-111.

Brooks, C. and Kreier, J. P. (1978). Role of the surface coat in in vitro attachment and phagocytosis of Plasmodium berghei by peritoneal macrophages. Infect. Immun., 20, 827-835.

Crossley, A. C. (1972). The ultrastructure and function of pericardial cells and other nephrocytes in an insect: Calliphora erythrocephala. Tissue and Cell, 4, 529-560.

Crossley, A. C. (1983). Nephrocytes and pericardial cells. In "Comprehensive Insect Physiology, Biochemistry and Pharmacology" (G. A. Kerkut and L. I. Gilbert, eds.), (In Press). Pergamon Press, London.

Cuénot, L. (1896). Etudes physiologiques sur les Orthoptères. *Arch. Biol.*, 14, 293-341.

Cuénot, L. (1905). L'organe phagocytaire des Crustacés Decapodes. *Arch. Zool. Exp. Gen.*, 4, 1-16.

Fontaine, C. T. and Lightner, D. V. (1974). Observations on the process of phagocytosis and elimination of carmine particles injected into the abdominal musculature of the white shrimp, *Penaeus setiferus*. *J. Invert. Pathol.*, 24, 141-148.

Hoffmann, J. A., Porte, A., and Joly, P. (1968a). Présence d'un tissu hématopoiétique au niveau du diaphragme dorsal *Locusta migratoria* (Orthoptére). *C. R. Acad. Sci.*, 266, 1882-1883.

Hoffmann, J. A., Stoeckel, M. E., Porte, A., and Joly, P. (1968b). Ultrastructure des hémocytes de *Locusta migratoria* (Orthoptère). *C. R. Acad. Sci.*, 266, 503-505.

Johnson, P. T. (1980). "Histology of the Blue Crab, *Callinectes sapidus*: A Model for the Decapoda." Prager, New York.

Lackie, A. M. (1981). The specificity of the serum agglutinins of *Periplaneta americana* and *Scistocerca gregaria* and its relationship to the insects' immune response. *J. Insect Physiol.*, 27, 139-143.

Lackie, A. M. (1983). Effect of substratum wettability and charge on adhesion *in vitro* by insect haemocytes. *J. Cell Sci.*, 181-190.

Martoja, R. and Martoja, M. (1967). "Initiation aux Techniques de l'Histologie Animale." Masson et Cie, Paris.

Rowley, A. F. and Ratcliffe, N. A. (1976). The granular cells of *Galleria mellonella* during clotting and phagocytic reactions *in vitro*. *Tissue Cell*, 8, 437-446.

Smith, V. J. and Ratcliffe, N. A. (1978). Host defense reactions of the shore crab *Carcinus maenas* (L), *in vitro*. *J. Mar. Biol. Assoc. U.K.*, 58, 367-379.

Smith, V. J. and Ratcliffe, N. A. (1980). Cellular defense of the shore crab, *Carcinus maenas: In vivo* hemocytic and histopathological responses to injected bacteria. *J. Invert. Pathol.*, 35, 65-74.

Vinson, S. B.   (1974).   The role of the foreign surface and
    female parasitoid secretions on the immune response of an
    insect.   *Parasitology*, 68, 27-33.

Zachary, D., Hoffmann, D., Hoffmann, J., and Porte, A.   (1981).
    Role of the reticulohemopoietic tissue of *Locusta
    migratoria* in the process of immunisation against *Bacillus
    thuringiensis*.   *Arch. Zool. Exp. Gen.*, 122, 55-63.

EVIDENCES FOR MOLECULAR SPECIFICITIES INVOLVED IN MOLLUSCAN
INFLAMMATION[1]

Thomas C. Cheng

Marine Biomedical Research Program
and Department of Anatomy (Cell Biology)
Medical University of South Carolina
Charleston, South Carolina 29412

I. INTRODUCTION

Cellular inflammatory response to nonself materials is a
universal phenomenon in the Animal Kingdom. To refresh our
memories, inflammation, as generally defined, is the reaction of
tissues to insult. The process is characterized by local heat,
swelling, redness, and pain. These characteristics have been

---

[1]This research was supported by a grant (PCM-8020884) from the
National Science Foundation.

defined as a result of observations on vertebrates, particularly mammals. As our knowledge of inflammation as revealed by studying invertebrates, especially molluscs, has progressed, it is becoming increasingly evident that these characteristics need not all be present, especially conspicuous swelling and redness. Also, I would like to advance the idea that nonspecific inflammatory response does not occur in molluscs or any other group of animals. In other words, there is some degree of specificity at the molecular level in all instances of cellular inflammation. The evidences for this statement are presented later. At this point, I wish to review in a conceptual manner the various phenomena which collectively comprise cellular response to insult in molluscs. Specifically, as depicted in Fig. 1, I intend to point out that cellular infiltration and subsequent events in molluscs, as in other groups of animals, involves (1) chemotactic attraction of reaction cells to the insulting agent, (2) surface recognition of self from nonself on the part of reaction cells and attachment of the nonself insulting agent to such cells, (3) endocytosis or encapsulation, and (4) intracelluar events leading to degradation and elimination of the insulting agent. As will become apparent at a later point, the reason for reviewing these phases of cellular reaction to challenge is that in terms of modern cell biology, there appears to be specific recognition sites involved at each phase, hence my thesis that nonspecific inflammatory response *per se* does not occur.

II. CHEMOTACTIC ATTRACTION

In view of the fact that my intent is to concentrate on the role of lysosomes in molluscan inflammation, only a few brief remarks is being made about attraction between reaction cells (primarily granulocytes) and the insulting agent.

That chemotaxis occurs between invading organisms and molluscan hemocytes was first demonstrated by Cheng et al. (1974). Subsequently, Schmid (1975) reported attraction of the hemocytes *Viviparus malleatus*, a pulmonate gastropod, to heat-killed *Staphylococcus aureus* and to N-acetyl-D-glucosamine in the presence of a naturally occurring hemagglutinin. Similarly, Cheng and Rudo (1976) have demonstrated the occurrence of chemotactic attraction between oyster (*Crassostrea virginica*) hemocytes and live *Micrococcus varians* and that *C. virginica* hemocytes are chemotactically attracted to several species of live Gram-positive and Gram-negative bacteria (Cheng and Howland, 1979). More recently Font (1980) has demonstrated attraction of *C. virginica* hemocytes to dead, but not living, cercariae of several species of marine trematodes.

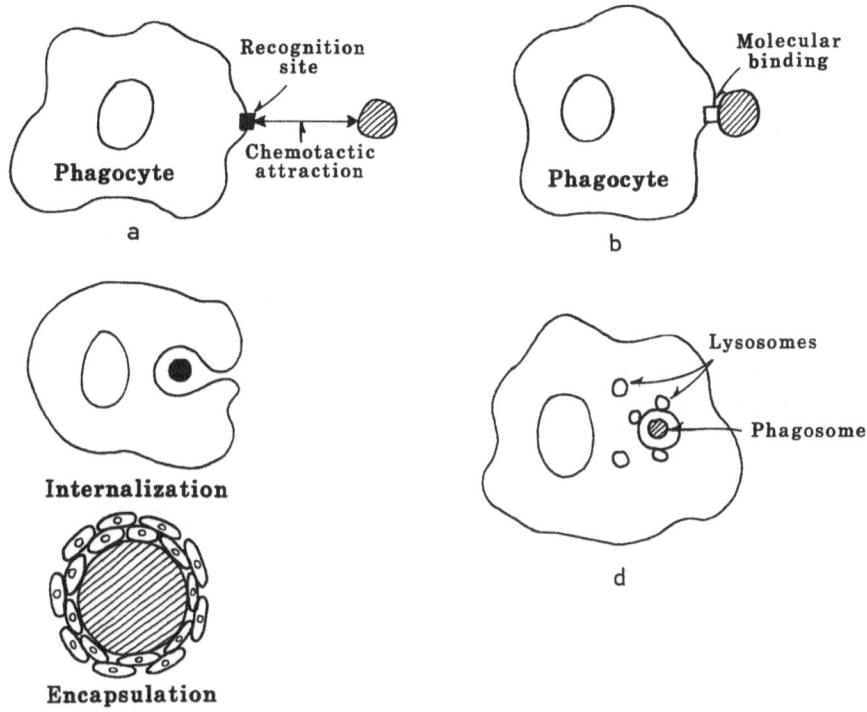

FIGURE 1.   Schematic drawings showing where molecular specifically
            do or may occur during cellular response to nonself
            materials in molluscs. A. Chemotaxis involving a
            recognition site. B. Molecular binding of the foreign
            material to the phagocyte. C. Internalization or encap-
            sulation. D. Fusion of primary lysosomes to phagosome.

Although the chemotactic signals emitted by *Bacillus
megaterium* and *Escherichia coli* have been elucidated (Howland and
Cheng, 1982), being proteins with molecular weights of approximately
10,000, the recognition site(s) on the molluscan phagocytes has yet
to be ascertained. Nevertheless, the ability of the host cells to
recognize the signal suggests the occurrence of such specific
site(s). This is schematically depicted in Fig. 2.

III. SURFACE RECOGNITION

The binding of nonself material to a molluscan phagocyte represents the second phase of cellular response. Again, in view of
what is known about cell surface binding in both vertebrates and

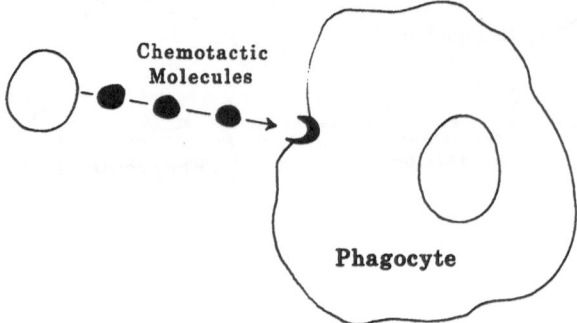

FIGURE 2.   Schematic drawing showing occurrence of a surface re-
            cognition site on the phagocyte specific for the chemo-
            tactic molecule(s).

invertebrates, this is usually not a random, nonspecific phenomenon.
Specifically, as has been reviewed earlier (Cheng et al., 1984;
Cheng, 1984) the currently most popular concept is that naturally
occurring lectins serve as bridges for binding the insulting agent
to molluscan phagocytes. Such lectins may occur in serum (Prowse
and Tate, 1969; Anderson and Good, 1976; Renwarantz, 1981), or are
integrally associated with the surface membrane of phagocytes
(Cheng et al., 1984; Vasta et al., 1984). The manifestation of
surface recognition mediated by lectins may be in the form of
simple attachment or encapsulation (Figs. 3,4). The major point I
wish to make is that there is specific binding during attachment
involving lectins.

IV.   ENDOCYTOSIS AND ENCAPSULATION

     Endocytosis of foreign materials by molluscan phagocytes has not
been studied in molecular terms, although at least one of three
morphologically distinguishable processes may be involved. Specifically, it has been reported that motile bacteria about to be endocytosed by granulocytes of *C. virginica* initially become adhered
to the surface of the molluscan cell, commonly on the surfaces of
filopodia (Bang, 1961). Subsequently, they are taken into the
ectoplasm by gliding along filopidia and become enclosed in a

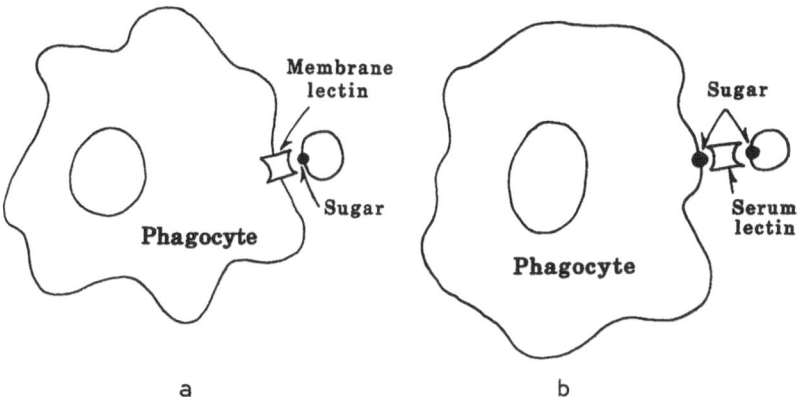

FIGURE 3.   Schematic drawings showing the mode of action of A, a
            membrane lectin and B, a serum lectin in binding the
            nonself body to the phagocyte.

phagosome (Fig. 5).   The second mechanism was discovered as a result of studying contact between certain bacteria (*Bacillus
megaterium*) and *C. virginica* granulocytes (Cheng, 1975).   This
mechanism involves the formation of invaginations on the cell
surface and the bacteria are endocytosed into vacuoles (Fig. 6).
No filopodia are involved.   The third endocytotic mechanism was
initially reported by Renwrantz et al.(1979).   They studied the
uptake or rat erythrocytes by *C. virginica* hemocytes.   They reported that funnel-shaped pseudopodia are formed and the erythrocytes
are endocytosed by gliding into a phagosome in the ectoplasm
(Fig. 7).   Subsequently, the occurrence of this uptake mechanism
by granulocytes of the gastropods *Biomphalaria glabrata* and
Bulinus truncatus have been also demonstrated (Schoenberg and
Cheng, 1980).

     The question that needs to be answered is: What are the
specific mechanisms governing which endocytotic process is manifested?   In any case, the point being made is that there appears
to be some specificity as how the foreign material is taken into
the phagocyte.

     Encapsulation is defined as the enveloping of a foreign body
which is dimensionally too large to be phagocytosed by host

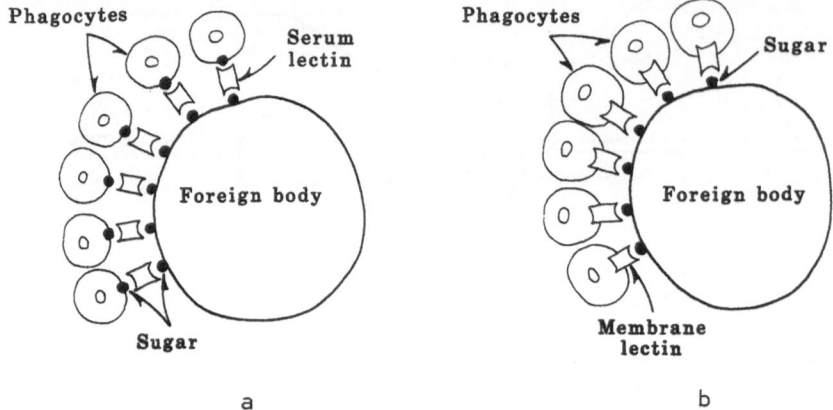

FIGURE 4.   Schematic drawings showing how A, a serum lectin and B,
            a membrane lectin could be involved in encapsulation.

hemocytes and/or fibers. This internal defense process, as stated
earlier, could be mediated by lectins. Thus, there is molecular
specificity involved.

V.   INTRACELLULAR DEGRADATION

     What is known about intracellular degradation of endocytosed
materials within molluscan phagocytes has been comprehensively reviewed (Cheng, 1981) and need not be repeated in detail. However,
from the standpoint of this presentation, it is important to note
that the release of lysosomal enzymes from primary lysosomes occurs
after fusion with compatible phagosomes. However, the degradation
of foreign particles and molecules is not restricted to secondary
lysosomes. This is the principal thrust of this presentation.

VI.   LYSOSOMAL HYDROLASES AND INFLAMMATION

     It has been stated earlier that the intracellular degradation
of endocytosed foreign materials is dependent on the fusion of
primary lysosomes and phagosomes followed by the release of

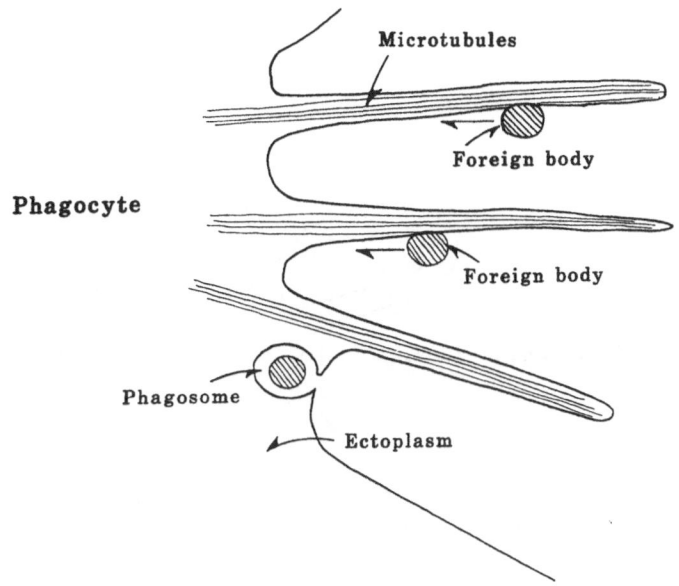

FIGURE 5.   Schematic drawing showing how a foreign body is endo-
            cytosed by the "Bang mechanism."

lysosomal hydrolases into secondary phagosomes. This, however, is
not the only role of lysosomal enzymes in molluscs.

As a result of ascertaining where lysozyme activity occurs in
molluscan hemolymph, it was found that this lysosomal hydrolase
occurs in both the serum and hemocytes of *C. virginica* and
Mercenaria mercenaria (Cheng and Rodrick, 1975). It was postulated
that the serum lysozyme had its origin in the lysosomes of granulocytes and was released when these organelles ruptured. Subsequently it was demonstrated that the lysosomal contents in granulocytes of the clam *M. mercenaria* are released by degranulation as
do those of mammalian macrophages. Also, it is now known that
degranulation, which represents the morphological basis for the
release of lysosomal enzymes, occurs rapidly in actively phagocytosing cells (Foley and Cheng, 1977).

Several lysosomal enzymes have been identified from the hemolymph of molluscs. Specifically, Rodrick and Cheng (1974) have
demonstrated lysozyme, acid phosphatase, β-glucuronidase, amylase,
and lipase in the freshwater gastropod *Biomphalaria glabrata* and

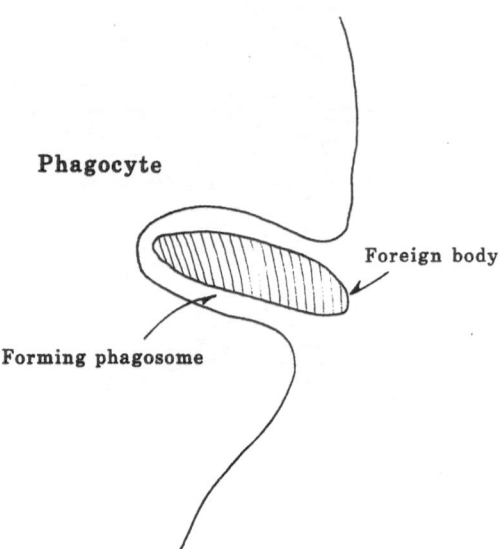

Phagocyte

Foreign body

Forming phagosome

FIGURE 6.   Schematic drawing showing the endocytosis of foreign
body by the formation of phagosome resulting from invagination of the phagocyte's surface membrane.

Cheng and Rodrick (1975) have reported these enzymes in the serum
and cells of *C. virginica* and *M. mercenaria*.   Furthermore, characterization of the lysozyme from *C. virginica* and *Mya arenaria* have
been carried out (Rodrick and Cheng, 1974; Cheng and Rodrick, 1974).
The lysozyme of *M. mercenaria* has also been characterized (Cheng,
1983) as has β-glucuronidase from *C. virginica* and *M. mercenaria*
(Cheng, 1976).

   The next advancement in our understanding of lysosomal enzymes
of molluscan hemocytes, primarily granulocytes, came about with
the demonstration on a quantitative basis that when *M. mercenaria*
hemocytes are induced to engage in phagocytosis by exposure to
Bacillus megaterium, there is enhanced release of lysozyme from
cells into serum (Cheng et al., 1975).   The same pattern holds
true *in vitro* and *in vivo* in the case of lipase from the granulocytic lysosomes of *Mya arenaria* challenged with *B. megaterium*
(Cheng and Yoshino, 1976a) and in the case of granulocytic
lysosomal lipase of *B. glabrata* challenged with live *B. megaterium*
(Cheng and Yoshino, 1976b).   However, when *B. glabrata* is

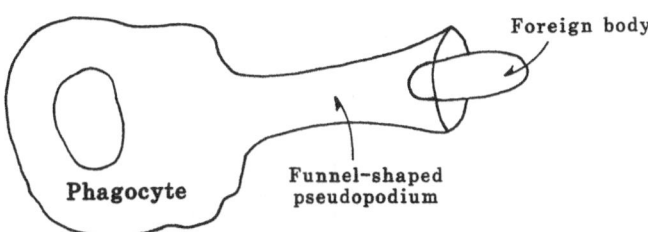

FIGURE 7.   Schematic drawing showing the formation of a funnel-
            shaped pseudopodium employed to endocytose a foreign
            body.

challenged with sonicated vegetative bacteria, there is no induced
elevation of lysosomal enzyme levels in either cells or serum.   Thus
some specificity apparently occurs since only live B. *megaterium*
induces the hypersynthesis of lipase within molluscan phagocytes
and the release of this lysosomal hydrolase into serum.

     Further evidence for the hypersynthesis of lysosomal enzymes
and their subsequent release into serum during active phagocytosis
on the part of molluscan hemocytes was contributed with the demonstration that *in vivo* challenge of B. *glabrata* with live B.
megaterium results in an elevation of this enzyme in hemocytes at
1 hr post-challenge and in serum and in serum at 2 and 4 hr post-
challenge (Cheng et al., 1977).   Additional studies demonstrating
hypersynthesis of lysosomal enzymes and their subsequent release
in activity phagocytosing molluscan hemocytes are available
(Cheng et al., 1978; Cheng and Butler, 1979).   These studies have
revealed that there is enhanced synthesis of aminopeptidase, lysozyme,
and acid phosphatase in actively phagocytosing cells of B. *glabrata*
and subsequent release of these hydrolases into serum.   The pattern
is now well established.

     It is of interest to note that in the case of inflammation
associated with gouty arthritis in humans, it has been shown that
the phenomenon is induced by the release of lysosomal enzymes by

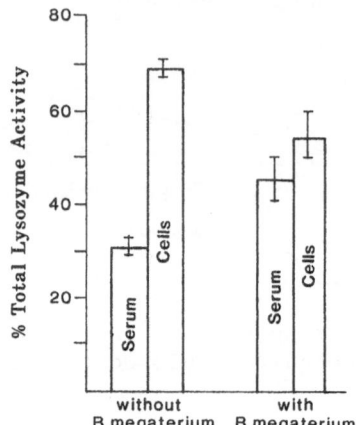

FIGURE 8.   Idiograms showing shift of intracellular lysozyme acti-
            vity to activity in the serum during phagocytosis of
            Bacillus megaterium by hemocytes (granulocytes) of
            Mercenaria mercenaria.

by polymorphonuclear leucocytes (Hoffstein and Weissman, 1975).
Other examples of inflammation caused by elevated levels of
lysosomal enzymes could also be cited. The patterns, as stated,
also holds true in molluscs challenged with nonself materials.

    In conclusion, I wish to reemphasize the role of specificity
in inflammation. It has been noted earlier that specific site(s)
probably occurs on molluscan phagocytes to permit the recognition
of chemotactic signals. Also, cell surface sugars or membrane
lectins bestow specificity relative to surface attachment, and circumstantial evidence suggest that the endocytotic process employed
for the internalization of nonself materials may be selectively
induced. The formation of secondary lysosomes as a part of the
intracellular degradation process is also governed by specificity
in that the primary lysosomes will not fuse with the phagosome
unless they are surfacially compatible at the molecular level.
Finally, the hypersynthesis of lysosomal enzymes and their subsequent release into serum, thus contributing to inflammation, is
selectively triggered, depending on the challenging material. Thus,
the concept is being advanced as a result of our studies involving
molluscan hemocytes that the cellular aspects of inflammation is

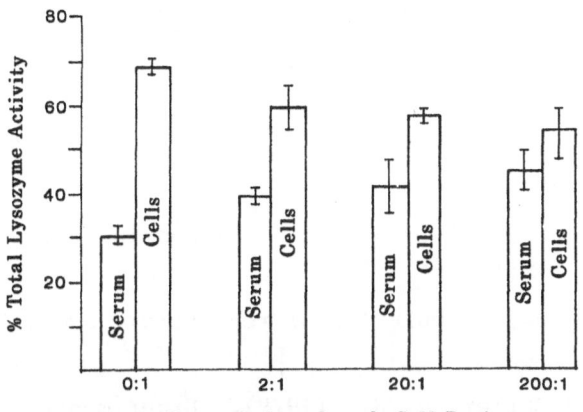

FIGURE 9.   Idiograms showing trend of increased release of lysozyme
            from hemocytes (granulocytes) of *Mercenaria mercenaria*
            into serum as the ratio of bacteria (*Bacillus megaterium*)
            to hemocytes is increased.

specific at the molecular level rather than what has been designated
as "nonspecific inflammatory response." Needless to state, a
great deal remains to be learned about the mechanisms governing
the various specificities.

VII.   REFERENCES

Anderson, R. S. and Good, R. A.   (1976).   Opsonic involvement in
    phagocytosis by mollusk hemocytes.   *J. Invert. Pathol.*, 27,57-64.

Bang, F. B.   (1961).   Reaction to injury in the oyster (*Crassostrea
    virginica*).   *Biol. Bull.*, 121, 57-68.

Cheng, T. C.   (1975).   Functional morphology and biochemistry of
    molluscan phagocytes.   *Ann. N.Y. Acad. Sci.*, 266, 343-379.

Cheng, T. C. (1976). Beta-glucuronidase from the serum and cells of *Mercenaria mercenaria* and *Crassostrea virginica* (Mollusca: Pelecypoda). *J. Invert. Pathol.*, 27, 125-128.

Cheng, T. C. (1981). Bivalves. *In* "Invertebrate Blood Cells 1" (N. A. Ratcliffe and A. F. Rowley, eds.) p. 233-300. Academic Press, London.

Cheng, T. C. (1983). The role of lysosomes in molluscan inflammation. *Am. Zool.*, 23, 129-144.

Cheng, T. C. (1984). Evolution of receptors. *Comp. Pathobiol.*, 5, 33-50.

Cheng, T. C. and Butler, M. S. (1979). Experimentally induced elevations of acid phosphatase activity in hemolymph of *Biomphalaria glabrata* (Mollusca). *J. Invert. Pathol.*, 34, 119-124.

Cheng, T. C. and Howland, K. H. (1979). Chemotactic attraction between hemocytes of the oyster, *Crassostrea virginica*, and bacteria. *J. Invert. Pathol.*, 33, 204-210.

Cheng, T. C. and Rodrick, G. E. (1974). Identification and characterization of lysozyme from the hemolymph of the soft-shelled clam *Mya arenaria*. *Biol. Bull.*, 147, 311-320.

Cheng, T. C. and Rodrick, G. E. (1975). Lysosomal and other enzymes in the hemolymph of *Crassostrea virginica* and *Mercenaria mercenaria*. *Comp. Biochem. Physiol.*, 52B, 443-447.

Cheng, T. C. and Rudo, B. M. (1976). Chemotactic attraction of *Crassostrea virginica* hemolymph cells to *Staphylococcus lactus*. *J. Invert. Pathol.*, 27, 137-139.

Cheng, T. C. and Yoshino, T. P. (1976). Lipase activity in the serum and hemolymph cells of the soft-shelled clam, *Mya arenaria*, during phagocytosis. *J. Invert. Pathol.*, 27, 243-245.

Cheng, T. C. and Yoshino, T. P. (1976). Lipase activity in the hemolymph of *Biomphalaria glabrata* (Mollusca) challenged with bacterial lipids. *J. Invert. Pathol.*, 28, 143-146.

Cheng, T. C., Cali, A., and Foley, D. A. (1974). Cellular reactions in marine pelecypods as a factor influencing endo-symbiosis. *In* "Symbiosis in the Sea" (W. B. Vernberg, ed.). p. 61-91. Univ. South Carolina Press, Columbia.

Cheng, T. C., Chorney, M. J., and Yoshino, T. P. (1977). Lysozyme-like activity in the hemolymph of *Biomphalaria glabrata* challenged with bacteria. *J. Invert. Pathol.*, 29, 170-174.

Cheng, T. C., Guida, V. G., and Gerhart, P. L. (1978). Amino-peptidase and lysozyme activity levels and serum protein concentrations in *Biomphalaria glabrata* (Mollusca) challenged with bacteria. *J. Invert. Pathol.*, 32, 297-302.

Cheng, T. C., Marchalonis, J. J., and Vasta, G. R. (1984). Role of molluscan lectins in recognition processes. *In* "Recognition Proteins, Receptors and Probes: Invertebrates". (E. Cohen, ed.) p. 1-15. Alan R. Liss, New York, N.Y.

Cheng, T. C., Rodrick, G. E., Foley, D. A., and Koehler, S. A. (1975). Release of lysozyme from hemolymph cells of *Mercenaria mercenaria* during phagocytosis. *J. Invert. Pathol.*, 25, 261-265.

Foley, D. A. and Cheng, T. C. (1977). Degranulation and other changes of molluscan granulocytes associated with phagocytosis. *J. Invert. Pathol.*, 29, 321-325.

Font, W. F. (1980). Effects of hemolymph of the American oyster, *Crassostrea virginica*, on marine cercariae. *J. Invert. Pathol.*, 36, 41-47.

Hoffstein, S. and Weissmann, G. (1975). Mechanism of lysosomal enzyme release. IV. Interaction of monosodium urate crystals with dogfish and human leukocytes. *Arthrit. Rheum.*, 18, 153-165.

Howland, K. H. and Cheng, T. C. (1982). Identification of bacterial chemoattractants for oyster (*Crassostrea virginica*) hemocytes. *J. Invert. Pathol.*, 39, 123-132.

Prowse, R. H. and Tate, N. N. (1969). *In vitro* phagocytosis by amoebocytes from the haemolymph of *Helix aspersa* (Müller). Evidence for opsonic factor(s) in serum. *Immunology*, 17, 437-443.

Renwrantz, L. (1981). *Helix pomatia*: Recognition and clearance of bacteria and foreign cells in an invertebrate. *In* "Aspects of Developmental and Comparative Immunology I" (J. B. Solomon, ed.) p. 133-138. Pergamon Press, N.Y.

Renwrantz, L., Yoshino, T., Cheng, T. C., and Auld, K. (1979). Size determination of hemocytes from the American oyster, *Crassostrea virginica*, and the description of a phagocytosis mechanism. *Jahrb. Zool. Abt. Physiol. Zoomorph.*, 83, 1-12.

Rodrick, G. E. and Cheng, T. C. (1974). Kinetic properties of lysozyme from *Crassostrea virginica* hemolymph. *J. Invert. Pathol.*, 24, 41-48.

Rodrick, G. E. and Cheng, T. C. (1974). Activities of selected hemolymph enzymes in *Biomphalaria glabrata* (Mollusca). *J. Invert. Pathol.*, 24, 374-375.

Schmid, L. S. (1975). Chemotaxis of hemocytes from the snail *Viviparus melleatus*. *J. Invert. Pathol.*, 25, 125-131.

Schoenberg, D. A. and Cheng, T. C. (1980). Phagocytic funnel-like pseudopodia in lectin-treated gastropod hemocytes. *J. Invert. Pathol.*, 36, 141-143.

Vasta, G. R., Cheng, T. C., and Marchalonis, J. J. (1984). A lectin on the hemocyte membrane of the oyster (*Crassostrea virginica*). *Cell Immunol.*, 88, 475-488.

Vasta, G. R., Sullivan, J. T., Cheng, T. C., Marchalonis, J. J., Warr, G. W. (1982). A cell membrane-associated lectin of the oyster hemocyte. *J. Invert. Pathol.*, 40, 367-377.

AN ELECTRON MICROSCOPE STUDY OF ENDOCYTOSIS MECHANISMS AND

SUBSEQUENT EVENTS IN *Mercenaria mercenaria* GRANULOCYTES

A. Mohandas

School of Environmental Studies
University of Cochin
Cochin-682 016
Kerala, India

I. INTRODUCTION

Foreign elements, molecules, or organisms naturally or experimentally introduced into naive molluscs are, although with exceptions, phagocytosed (see reviews by Cheng, 1975, 1981). Of the
two categories of hemocytes common to all bivalves, hyalinocytes
and granulocytes (Cheng, 1981), the latter in *Crassostrea virginica*
and *Mercenaria mercenaria* are found to be the most active from the
standpoint of phagocytosis (Foley and Cheng, 1975). There are
three ways, involving semi-permanent filopodia, classical endocytosis, and afunnel-like pseudopodia, by which bacteria are engulfed
by hemocytes. The uptake mechanism has been studied at the
electron microscope level in *Crassostrea gigas* (Ruddell, 1971;
Feng et al., 1977), *C. virginica* (Cheng and Cali, 1974; Cheng,

143

1975), and *Mytilus coruscus* (Feng et al., 1977). Furthermore, as a
rule, such phagocytosed materials, if digestible, are degraded by
intracellularly, and some are removed when cells containing such
materials migrate to the exterior through epithelial borders (Cheng,
1977a). Since granulocytes are more phagocytic than hyalinocytes,
and one of the major differences between these two types of cells
is the occurrence of large numbers of cytoplasmic granules in
granulocytes, further studies to resolve the nature of these granules
revealed that those in *C. virginica* are electron-lucid vesicles each
of which possesses a complex wall (Feng et al., 1971; Cheng and
Cali, 1974; Cheng et al., 1974; Cheng, 1975). However, the granules
of *M. mercenaria* granulocytes, which are membrane-bound vesicles
containing a homogeneously electron-dense substance (Cheng, 1975;
Cheng and Foley, 1975), have subsequently been demonstrated to be
true lysosomes (Yoshino and Cheng, 1976). Furthermore, it has been
established that when challenged with foreign materials, there is
hypersynthesis of certain lysosomal hydrolases within the hemocytes
and their subsequent release into serum (Cheng et al., 1975; Cheng
and Butler, 1979; Cheng and Mohandas, 1985; Mohandas and Cheng,
1985). These lysosomal hydrolases, which are associated with
intracellular degradation (Cheng and Cali, 1974; Cheng et al., 1974),
have also been demonstrated to have antimicrobial properties
(McDade and Tripp, 1967; Cheng, 1978).

By employing *M. mercenaria* granulocytes as the model, it has
been demonstrated biochemically and semiquantitatively that degranulation in molluscan granulocytes is the morphological reflection
of the release of enzymes into the medium (Cheng, 1975; Cheng et al.,
1975; Foley and Cheng, 1977). A subsequent scanning electron microscope study, designed to demonstrate the mechanism of lysosomal
enzyme release from *M. mercenaria* granulocytes at 1 and 2 hr post-
challenge with heat-killed *Bacillus megaterium in vitro*, has revealed that lysosomes protrude from the plasma membrane of the cells
and bud off into serum (Mohandas et al., 1985). Since it is known
that granulocyte's plasma membrane is not radically disrupted during
degranulation (Cheng, 1975; Cheng et al., 1975), it was hypothesized
that each lysosome discharged into serum is bound by two membranes;
the inner lysosomal membrane and the outer granulocyte plasma membrane (Mohandas et al., 1985).

Since there are some differences between the cytoplasmic
granules of *M. mercenaria* and those of *C. virginica*, and since the
fate of bacteria phagocytosed by granulocytes of *C. virginica* has
been studied by employing transmission electron microscopy (TEM)
(Cheng and Cali, 1974), it was thought worthwhile to employ TEM to
ascertain the fate of bacteria phagocytosed by granulocytes of
M. mercenaria.

The present TEM study on *M. mercenaria* granulocytes was designed to reveal (1) the uptake mechanism of heat-killed *B. megaterium*, (2) the fate phagocytosed bacteria, and (3) the process of degranulation.

II. MATERIALS AND METHODS

The clams, *M. mercenaria*, used in the present study were obtained from a commercial source in Charleston, South Carolina, but orginiated in that state. For 8-10 weeks prior to use, they were maintained in recirculating artificial sea water (20º/oo salinity, 22º±1ºC) tanks fortified with the diatom *Thallassiosira* sp.

A millipore (0.22 µm) filtered, sterile sea water (salinity 20º/oo) suspension of heat-killed *B. megaterium* at a concentration of 4 x 10⁶ bacteria/ml was employed as the challenging agent. This strain of bacteria (ATCC 14581) was obtained from the American Type Culture Collection, Rockville, Maryland, and has been maintained in culture in Dr. Cheng's laboratory for over 5 years in trypticase soy broth, involving weekly transfer.

TEM. Hemolymph samples were drawn from the heart of ten *M. mercenaria* by employing a 5-ml syringe fitted to a 18-gauge hypodermic needle, emptied into 15-ml centrifuge tubes, and used within 2 min. Specifically, a 0.5 ml sample of whole hemolymph from each clam was discharged into a 4-ml siliconized vial to which a 0.5 ml of bacterial suspension was added immediately. The preparation was thoroughly mixed, and the vials were incubated in humidified chambers at 22º±1ºC. At 1 and 2 hr post-challenge, the hemocyte pellets, concentrated by centrifugation at 1060*g* for 10 min, were fixed in ice-cold 2% gluteraldehyde, pH 7.3, and buffered with 0.1M sodium cacodylate for 2 hr. After rinsing four times with cold buffer during 1 hr, the pellets were post-fixed for 2 hr in ice-cold 2% osmium tetroxide buffered to pH 7.3 with 0.1M sodium cacodylate. They were dehydrated in a graded ethanol series, embedded in Spurr's epon, and cold sections cut with a diamond knife (DuPont) on a Sorvall ultramicrotome were picked upon 200-mesh copper grids. Following staining with uranyl acetate and lead citrate, 4 min each, the sections were examined in a Hitachi HU-11E-2 electron microscope operated at 75kV.

III. RESULTS

Based on the examination of several hundred granulocytes of *M. mercenaria* challenged *in vitro* with heat-killed *B. megaterium*, the following observations have been made.

In the first type of uptake mechanism, which is comparatively
rare, filopods of each granulocyte, supported by microtubules
(Fig. 1), extend toward the bacteria, adhere to them (Fig. 2), and
the bacteria are subsequently endocytosed into the ectoplasm. More
precisely, with the formation of filopods, the spaces between them
are gradually occupied by out-flowing ectoplasm, which forms a
web between adjacent pseudopods. Such ectoplasmic webs contain
a large number of microtubules (Fig. 3) and it is in such webs
that endocytotic vesicles are formed.

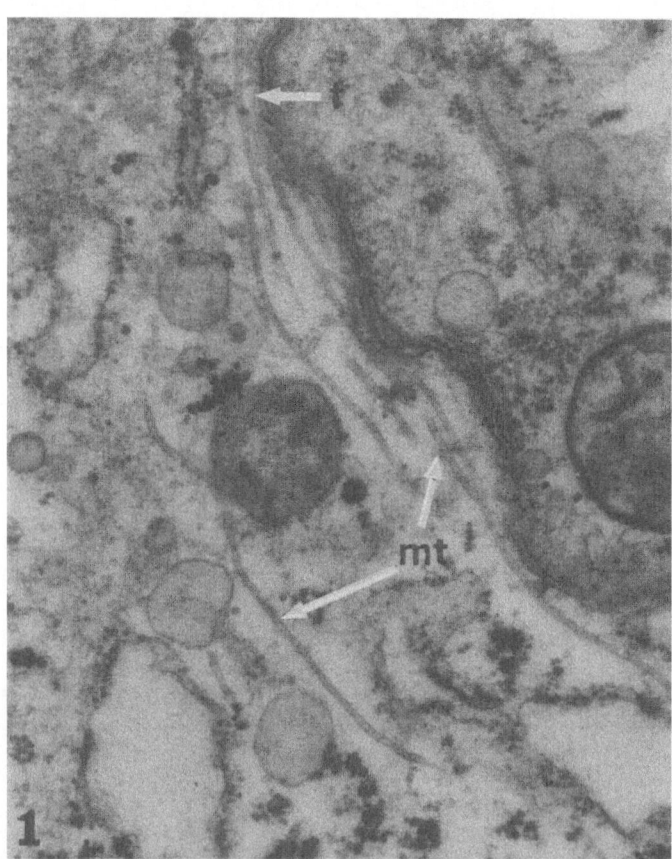

FIGURE 1.   Electron micrograph of portion of *Mercenaria mercenaria*
             granulocyte at 1 hr post-challenge with *Bacillus
             megaterium* showing filopod (f) and microtubules (mt).
             X46,360.

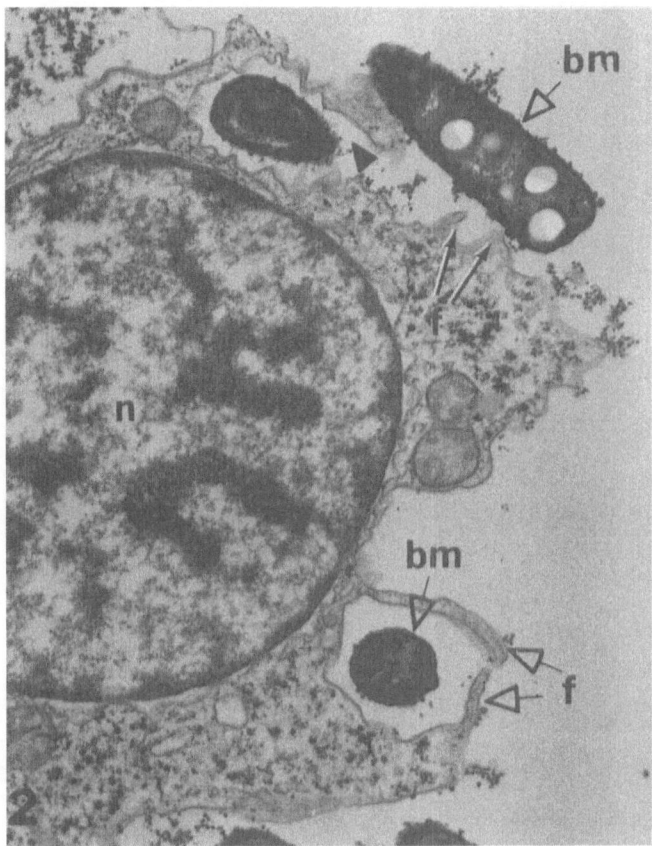

FIGURE 2.   Electron micrograph of portion of *Mercenaria mercenaria*
            granulocyte at 1 hr post-challenge with *Bacillus*
            megaterium showing filopods (f) extending towards
            bacterium (bm) and sticking to it. Another bacterium
            is taken in by filopods, and a phagosome is being formed
            (thick arrow head).   n=granulocyte nucleus. X13,664.

     In the second uptake mechanism, which occurs more frequently
and does not involve filopods, after contact has been achieved
between a granulocyte and bacteria, invaginations on the surface
of the phagocyte develop and the bacteria are taken into the
endocytotic vesicles (Fig. 3). The channel leading into each
vesicle is also lined by arrays of microtubules (Fig. 3). Such a
primary phagosome usually includes only one bacterium but it is not
uncommon to find more than one bacterium within the phagosome
(Fig. 4). Such a condition is the result of fusion of adjacent

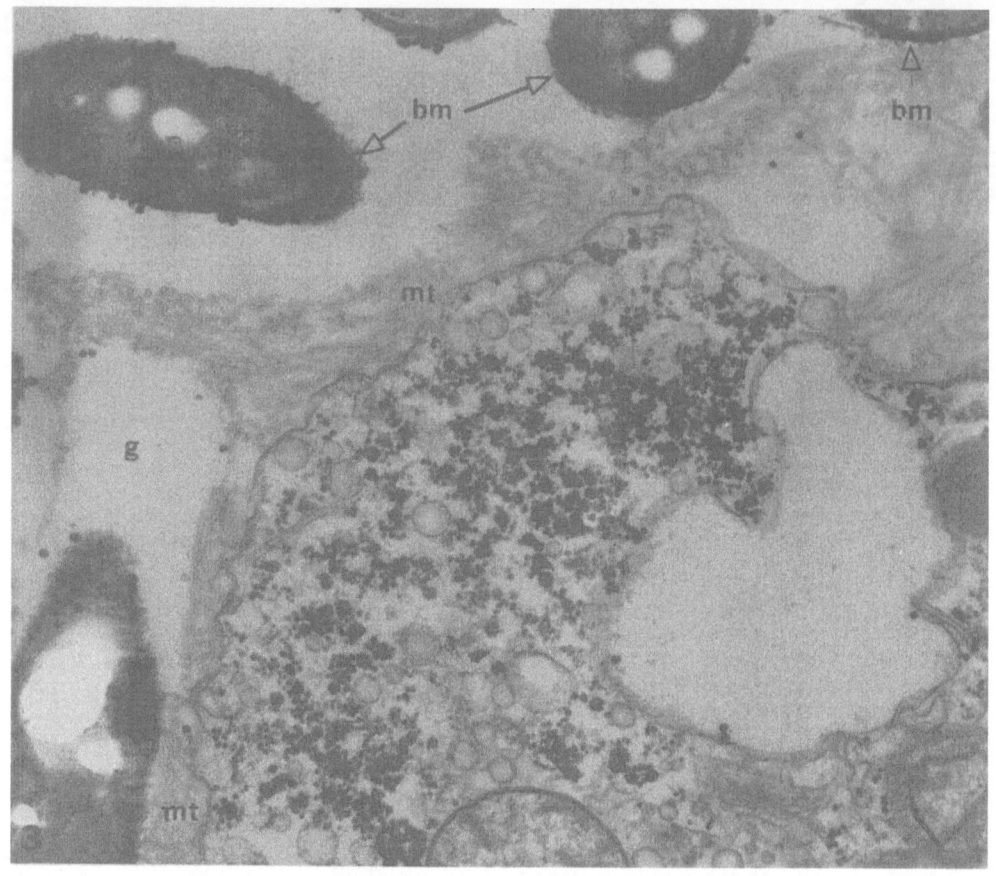

FIGURE 3.   Electron micrograph of *Mercenaria mercenaria* granulocyte
at 1 hr post-challenge with *Bacillus megaterium* showing
microtubules (mt) in ectoplasm, and also along the groove
(g) leading to endocytotic vesicle. Note bacteria (bm)
in contact with microtubules (mt). X33,400.

phagosomes. That the bacteria are digested within these phagosomes
is suggested by the disorganized and degraded appearance of their
integrity (Fig. 4) as well as the appearance of digestive lamellae
along their periphery (Fig. 5). These intraphagosomal lamellae are
comprised of 5-6 electron-dense membranous rings, with a lucid zone
between the rings. Each membrane appears to be a typical unit
membrane, i.e., with an electron-lucid zone situated between two
dense zones.

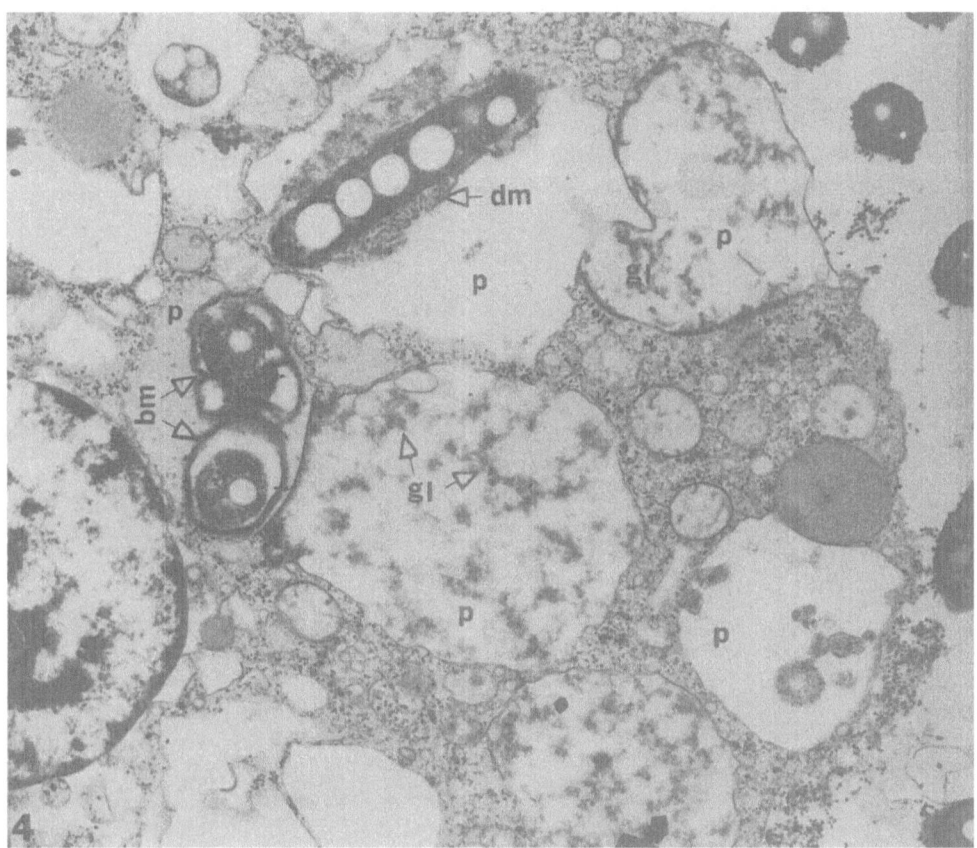

FIGURE 4. Electron micrograph of portion of *Mercenaria mercenaria*
           granulocyte at 2 hr post-challenge with *Bacillus*
           megaterium showing bacteria (bm), bacterial degraded
           material (dm), and glycogen (gl) in primary phagosomes
           (p). Note two bacteria in one phagosome. X13,350.

     Degradation of the bacteria is completed within phagosomes.
In some phagosomes degraded materials are apparent (Fig. 4). These
appear to represent an intermediate products in the intraphagosomal
synthesis of glycogen. In some other phagosomes glycogen granules
occur (Fig. 4). Such glycogen granules have not been observed to be
deposited free in the cytoplasm due to the fragmentation of the
phagosomal wall. On the contrary, the occurrence of phagosomes,
containing clumps of glycogen granules and nondigestible materials,
close to the plasma membrane with sprout-like extensions merging

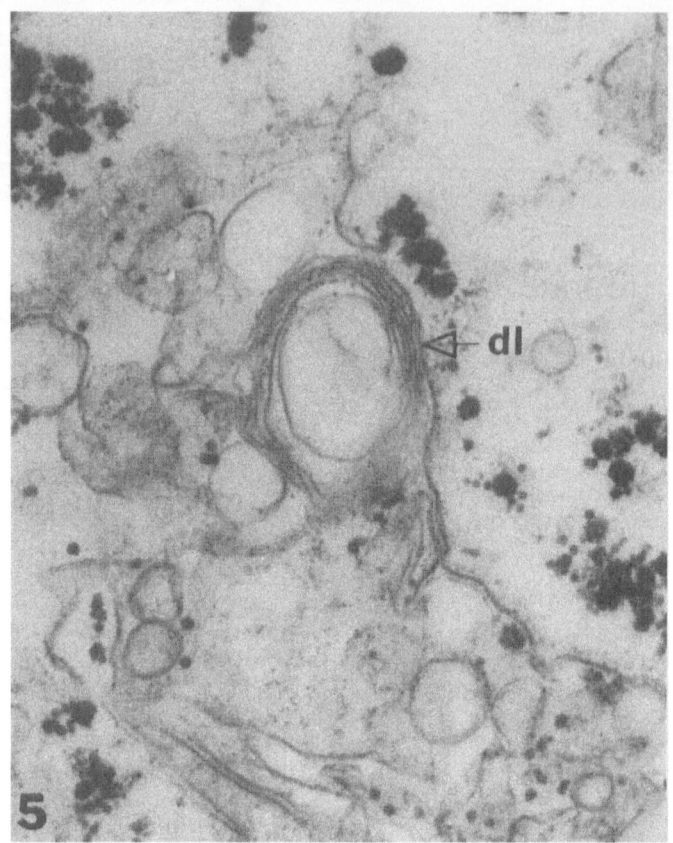

FIGURE 5.   Electron micrograph of portion of *Mercenaria mercenaria*
            granulocyte at 1 hr post-challenge with *Bacillus
            megaterium* showing digestive lamellae (dl) around
            bacterium.   X13,664.

with the granulocytes plasma membrane suggests the process of expul-
sion of these glycogen granules and nondigestible materials from
phagosomes to the exterior of the granulocyte, obviously into the
serum.

     Cytochemically and functionally, the lysosomes form a hetero-
genous population and appear as electron-dense or electron-lucid
vesicles (Fig. 6). These vesicles are found throughout the endo-
plasma and also close to the plasma membrane. During degranulation,
which is a normal process but is greatly enhanced when stimulated by

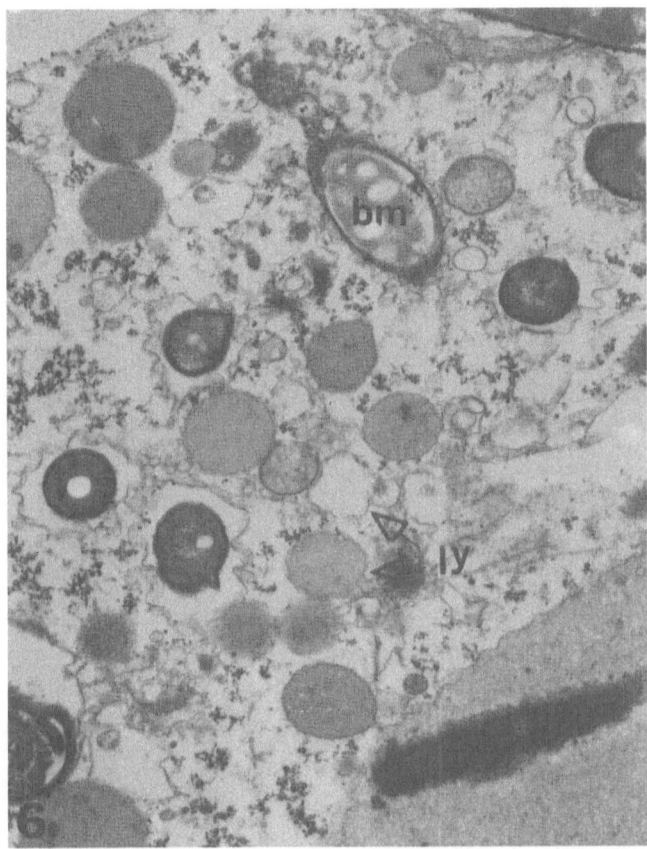

FIGURE 6.   Electron micrograph of portion of *Mercenaria mercenaria*
            granulocyte at 1 hr post–challenge with *Bacillus*
            megaterium showing electron-dense and electron-lucid
            lysosomes (ly) and bacterium (bm). X13,664.

phagocytosis of *B. megaterium*, lysosomes merge with the plasma
membrane of the granulocyte (Fig. 7), protrude on the granulocyte's
surface (Fig. 7), and are released into the serum (Fig. 8). Each
released lysosome is bound by two membranes: the inner lysosomal
membrane and the outer granulocyte plasma membrane (Fig. 8). Some
of the lysosomes within the granulocytes contain a relatively large
amount of indigestible or partially digestible remnants of bacterial
breakdown products. These may be considered as residual bodies
(Fig. 7).

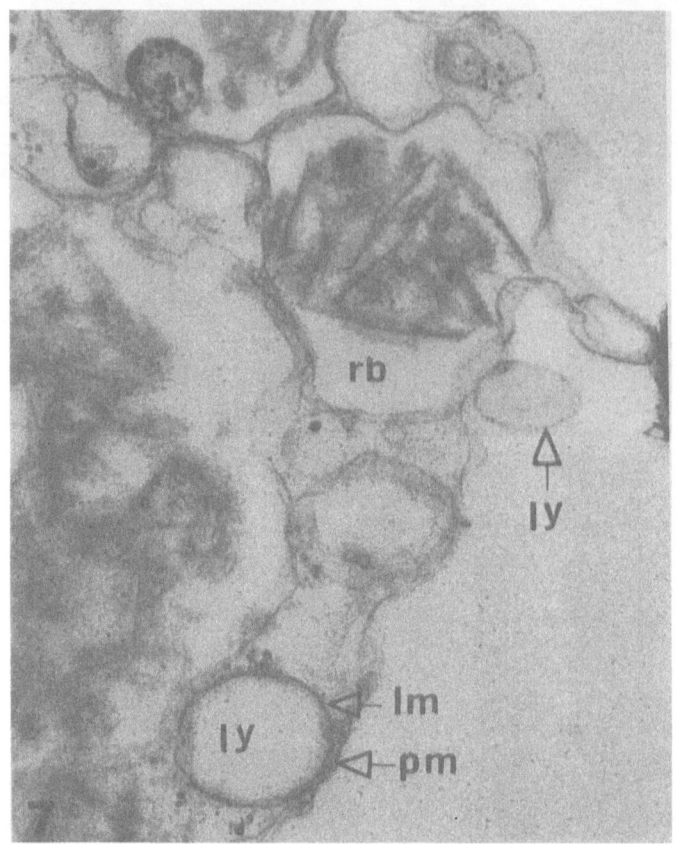

FIGURE 7.   Electron micrograph of portion of *Mercenaria mercenaria*
            granulocyte at 2 hr post-challenge with *Bacillus*
            megaterium showing lysosome (ly) merging with granulocyte
            plasma membrane (pm) and released lysosome.   lm=
            lysosomal membrane, rb=residual body.   X46,360.

IV. DISCUSSION

     Three types of uptake mechanisms involving filopods, endocyto-
sis, and funnel-like pseudopods have been reported in molluscs
(Cheng, 1981).   Bang (1961) has reported that motile bacteria about
to be phagocytosed by granulocytes of *C. virginica* initially stick
to the molluscan cell surface, generally to the surface of a filopod, and are taken into the ectoplasm and become enclosed in a
phagosome. In the present study these features also have been
observed in *M. mercenaria* granulocytes, although less frequently.

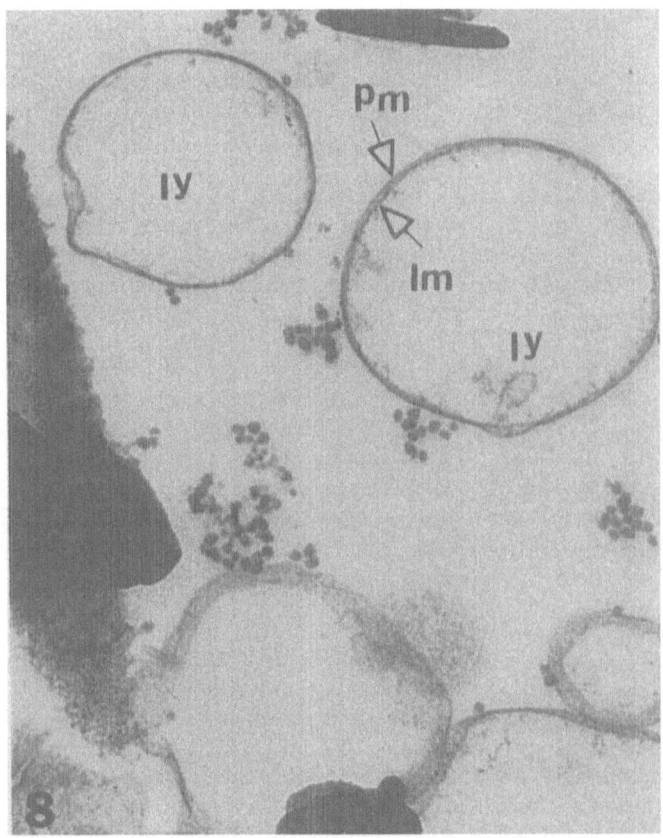

FIGURE 8.   Electron micrograph of portion of *Mercenaria mercenaria*
            granulocyte at 2 hr post-challenge with *Bacillus*
            megaterium showing lysosomes (ly) released into serum.
            lm=lysosomal membrane, pm=granulocyte plasma membrane.
            X46,360.

     It has also been observed in the present study that the filo-
pods include microtubules which extend internally into the endoplasm.
Earlier, Foley and Cheng (1972) have reported that in *C. virginica*
granulocytes, each filopod is supported by a rib-like structure
oriented along its length and extending internally into the endoplasm. Subsequently, these rib-like structures have been demonstrated to be comprised of a fascicle of microtubules (Cheng, 1975).
It is believed that these organelles aid in maintaining the
rigidity of the filopods (Cheng, 1975).

It has been reported above that the ectoplasm that forms a
web between adjacent pseudopods also contains a rich array of
microtubules. It is noted that Bang (1961) has reported similar
ectoplasmic web between adjacent pseudopods in *C. virginica*
hemocytes, and has suggested that this cytoplasmic web serves
as a trap and the uptake of bacterial cells is effected when they
are trapped in the web. The role of microtubules in the ectoplasm
is also assumed to be supportive because besides giving additional
support to the filopods from outside to maintain rigidity, these
organelles provide support to the ectoplasmic web itself to maintain
rigidity. The longitudinal and latitudinal arrangement of these
microtubules may be to restrict the lateral shifting and marked
bending of pseudopods.

The main uptake mechanism that has been observed in *M.
mercenaria* granulocytes is the one involving classical endocytosis.
It is believed that the microtubules also are involved in trapping
heat-killed bacteria into the ectoplasmic web, and help them
glide through the channel into the primary phagosomes. The presence of bacteria in the web, microtubules all along the channel,
and bacteria inside the channel all are supportive of this contention.

It is noted that Cheng and Howland (1982) have demonstrated
that chemotaxis between *C. virginica* hemocytes and live *B.
megaterium* is significantly reduced when the cells are pretreated
with colchicine and cytochalasin B, and indicated that hemocytes of
C. virginica require an intact cytoskeletal system to respond
chemotactically to the bacteria. The findings by Rodrick and
Ulrich (1984) are in line with those of Cheng and Howland (1982).
Colchicine inhibits the assembly of microtubules by binding to
monomeric tubulin and thus prevents its polymerization into microtubules (Bryan, 1974). Malech et al. (1977) have demonstrated that
in mammalian leukocytes colchicine causes a dramatic loss in cell
orientation in a chemotactic gradient that is related to a decrease
in the number of microtubules and concluded that the microtubule
apparatus stabilizes the cell during chemotaxis and contributes to
the vector of locomotion.

An answer to the question as to why more heat-killed *B.
megaterium* are taken in by endocytosis has been provided by Cheng
(1975). His contention is that bacteria that are taken into host
cells by endocytosis have already been altered by enzymatic action.
It has already been demonstrated that serum of *Biomphalaria glabrata*
which had previously been challenged with heat-killed *B. megaterium*
as well as pure lysozyme cause erosion of cell wall and disruption
of cellular inclusions of live *B. megaterium* (Cheng, 1977b, 1978).
Moreover, our previous SEM studies on *M. mercenaria* granulocytes
challenged with heat-killed *B. megaterium* and examined at 1 and 2

hr post-challenge (Mohandas et al., 1985) as well as the present
study have revealed that the bacteria that are endocytosed by
granulocytes have altered body surface resulting from lysosomal
enzyme action. It may be that partial degradation of bacterial
cell wall, at least of certain species, may be a prerequisite for
endocytosis by granulocytes.

The endocytosed bacteria are taken into large membrane-lined
phagosomes. Further degradation of the bacteria is initiated and
completed within these organelles. The presence of degraded
bacteria, formation of digestive lamellae around these, and clumps
of glycogen granules and nondigestible materials within phagosomes
indicate that the entire process of bacterial degradation is
completed within these organelles. What had been designated as
"secondary phagosomes" by Cheng and Cali (1974) appear to be absent,
and phagosomal walls do not appear to break up so as to release
glycogen into the cytoplasm as is apparently the case in *C.
virginica* (see Cheng and Cali, 1974). Interestingly, in *C.
virginica* granulocytes the partially degraded bacteria are trans-
ferred to "secondary phagosomes" (a term which is probably inappropriate an opinion with which Dr. Cheng agrees) where further
degradation occurs and subsequently the phagosomal wall breaks up
resulting in the deposition of clumps of glycogen in the cytoplasm
(Cheng and Cali, 1974; Cheng, 1975). Later, glycogen is discharged
into serum, enveloped by the plasma membrane of the granulocyte
(Cheng and Cali, 1974; Cheng, 1975).

It is noted that Cheng (1975) has reported a significant rise
in hemolymph glycogen in *C. virginica* at 2 hr post-challenge to
Escherichia coli in vivo, and Rodrick and Ulrich (1984) have obtained similar results in *Mercenaria campechiensis, C. virginica*,
and *Anadara ovalis* at 1 hr post-challenge with *E. coli* and *Vibrio
anguillarum in vivo*. By employing ^{14}C-labeled *B. megaterium* as
the challenging agent, Cheng (1977a) also has demonstrated that
the degradation of phagocytosed bacteria in *C. virginica* hemocytes
leads to the synthesis of glycogen from sugar of bacterial origin
and its eventual release from phagocytes. In *C. virginica* as well
as in *M. mercenaria*, the indigestible components are expelled from
phagosomes to the exterior of the phagocyte.

In the present study, concentric lamellae have been found in
phagosomes surrounding partially digested bacteria similar to the
condition in *C. virginica* granulocytes (Cheng and Cali, 1974;
Cheng, 1975). These lamellae have been reported to be associated
with intracellular digestion in a variety of cells (Swift and
Hruban, 1964; Hohl, 1965). Hohl (1965) has expressed the opinion
that these organelles are synthesized from products resulting from
digested foreign particles, and Cheng and Cali (1974) consider it
highly likely that the concentric lamellae found within phagosomes

of *C. virginica* granulocytes may also be synthesized from breakdown products of microorganisms phagocytosed by the hemocytes.

Under normal conditions, lysosomes in *M. mercenaria* granulocytes are mostly scattered throughout the endoplasm and are only
rarely observed in the ectoplasm (Cheng and Foley, 1975). The present study, however, has revealed that as a prelude to the process
of degranulation subsequent to challenge with heat-killed *B.
megaterium*, lysosomes migrate towards the granulocyte's periphery
to abut the overlying granulocyte plasma membrane followed by pro-
trusion beyond the granulocyte surface. Each of these lysosomes,
i.e., those that protrude from the plasma membrane of the cell as
well as those that bud off into serum, is bound by two membranes:
the inner lysosomal and the outer granulocyte membrane.

Cytochemically and functionally, the lysosomes form a heterogenous population (Dean, 1977; Schellens et al., 1977). In *M.
mercenaria* granulocytes, they appear as electron-dense and electron-
lucid vesicles, and the finding that not all of the lysosomes in
each cell include acid phosphatase suggests a nonsynchronized
chemical cycle occurring within this organelle or the lysosomes
represent a chemically heterogenous population (Yoshino and Cheng,
1976). Lysosomes even in single cell types are quite variable in
their enzymatic constitution (Dean, 1977). Also, this organelle
differs morphologically in the granulocyte of *M. mercenaria* (Cheng
and Foley, 1975), and this heterogeneity in size and shape reflects
the divergent functional activities of lysosomes in different types
of cells (Schellens et al., 1977).

Although the process of degranulation has been elucidated in
the granulocytes of *M. mercenaria*, the mechanism(s) responsible
for the detachment of lysosomes from the granulocyte plasma membrane and the subsequent lysis of lysosomes in the serum so as to
release the enclosed hydrolases has yet to be investigated. It is
possible, as suggested earlier (Mohandas et al., 1985), that an
autolytic mechanism may be operative in the release of hydrolases
from the double-membraned lysosomes. It may also be due to
difference in osmotic pressure within the released lysosomes and in
the serum caused by calcium influx. In the mammalian circulatory
system, calcium is known to cause sudden change in osmotic pressure
due to the formation of colloidal calcium phosphate (Venugopal and
Luckey, 1978).

V.   CONCLUSIONS

Although the process of degranulation has been elucidated in
M. mercenaria granulocytes, it has yet to be ascertained whether the
same mechanism is operative in other molluscan species. In a recent
review on hemocytes of bivalves by Cheng (1981), the functional

basis for recognizing two categories of cells, hyalinocytes and granulocytes, has been presented. Earlier, Renwrantz et al. (1979) have demonstrated that, based on their dimensions, the granulocytes of *C. virginica* can be subdivided into three subpopulations. Later, by employing density gradient centrifugation and identifying the surface receptors of *C. virginica* hemocytes by employing lectins, Cheng et al. (1980) have demonstrated that the subpopulations of *C. virginica* granulocytes can be subdivided into five subpopulations. Results of additional studies on hemocyte lectin receptors and E-rosetting properties, and the recent application of monoclonal antibodies as molecular probes of molluscan hemocyte surface antigens suggest that the population of circulating molluscan hemocytes may actually be composed of molecularly and antigenically distinct cell subpopulations potentially capable of some degree of functional compartmentalization (Yoshino and Granath, 1985). The implication of these findings relative to the recognition of exogenous stimulation is not yet known, although it is also possible that the different subpopulations of granulocytes are qualitatively and/or quantitiatively differentially stimulated by exogenous factors (Cheng, 1983).

It is, however, not yet known whether all the molluscan species have different functional subpopulations of granulocytes, and if so, whether the different functions attributed to molluscan granulocytes are being performed by different subpopulations of cells. Moreover, it is also not yet clear whether the different subpopulations of granulocytes show different lysosomal enzymes activities, especially when it is known that lysosomes even in a single cell type are quite variable in their enzymatic constitution. It has yet to be demonstrated whether a particular challenging agent evokes the same or different stimulus in the different subpopulations of granulocytes resulting in the release of one or several types of lysosomal acid hydrolases. Finally, it is difficult at this stage to say definitely whether the chemically heterogenous populations of lysosomes showing acid phosphatase activity in the granulocytes of *M. mercenaria* (Yoshino and Cheng, 1976) are constituents of a particular subpopulation only or whether these are constituents of all subpopulations but exhibiting the same biochemical activity upon challenge with a suitable antigenic agent.

VI. ACKNOWLEDGEMENTS

The studies reported herein were conducted while the author was a Visiting Research Scientist in the laboratory of Dr. Thomas C. Cheng at the Marine Biomedical Research Program, Medical University of South Carolina, Charleston, South Carolina, U.S.A. The partial support from the Council for International Exchange of Scholars is gratefully acknowledged. Research support was also

provided by a grant (PCM–8208016) from the National Science
Foundation and a contract (DE-ASO9–83ER60132) from the U.S.
Department of Energy to Dr. Cheng.

VII. REFERENCES

Bang, F. B. (1961). Reaction to injury in the oyster (*Crassostrea
virginica*). *Biol. Bull.*, 121, 57–68.

Bryan, J. (1974). Microtubules. *Bioscience*, 24, 701–711.

Cheng, T. C. (1975). Functional morphology and biochemistry of
molluscan phagocytes. *Ann. N.Y. Acad. Sci.*, 266, 343–379.

Cheng, T. C. (1977a). Biochemical and ultrastructural evidence
for the double role of phagocytosis in molluscs: defense and
nutrition. *Comp. Pathobiol.*, 3, 21–30.

Cheng, T. C. (1977b). The role of hemocytic hydrolases in the
defense of molluscs against invading parasites. *Haliotis*,
8, 193–209.

Cheng, T. C. (1978). The role of lysosomal hydrolases in molluscan
cellular response to immunologic challenge. *Comp. Pathobiol.*,
4, 59–71.

Cheng, T. C. (1981). Bivalves. *In* "Invertebrate Blood Cells",
N. A. Ratcliffe and A. F. Rowley (eds.). Academic Press,
London. pp. 233–300.

Cheng, T. C. (1983). The role of lysosomes in molluscan inflammation. *Amer. Zool.*, 23, 129–144.

Cheng, T. C. and Butler, M. S. (1979). Experimentally induced
elevations in acid phosphatase activity in hemolymph of
Biomphalaria glabrata (Mollusca). *J. Invertebr. Pathol.*,
34, 119–124.

Cheng, T. C. and Cali, A. (1974). An electron microscope study
of the fate of bacteria phagocytized by granulocytes of
Crassostrea virginica. *Cont. Topics. Immunobiol.*, 4, 25–35.

Cheng, T. C. and Foley, D. A. (1975). Hemolymph cells of the
bivalve mollusc *Mercenaria mercenaria*: An electron microscopical study. *J. Invertebr. Pathol.*, 341–351.

Cheng, T. C. and Howland, K. H. (1982). Effects of colchicine
and cytochalasin B on chemotaxis of oyster *Crassostrea
virginica)* hemocytes. *J. Invertebr. Pathol.*, 40, 150–152.

Cheng, T. C. and Mohandas, A. (1985). Effect of high dosages of bacterial challenge on acid phosphatase release from *Biomphalaria glabrata* hemocytes. *J. Invertebr. Pathol.*, 45, 236-241.

Cheng, T. C., Cali, A., and Foley, D. A. (1974). Cellular reactions in marine pelecypods as a factor influencing endosymbioses. *In* "Symbiosis in the Sea", W. B. Vernberg (ed.). University of South Carolina Press, Columbia, South Carolina. pp. 61-91.

Cheng, T. C., Rodrick, G. E., Foley, D. A., and Koehler, S. A. (1975). Release of lysozyme from hemolymph cells of *Mercenaria mercenaria* during phagocytosis. *J. Invertebr. Pathol.*, 25, 261-265.

Cheng, T. C., Huang, J. W., Karadogan, H., Renwrantz, L. R., and Yoshino, T. P. (1980). Separation of oyster hemocytes by density gradient centrifugation and identification of their surface receptors. *J. Invertebr. Pathol.*, 36, 35-40.

Dean, R. T. (1977). "Lysosomes", The Institute of Biology's Studies in Biology, No. 84, Edward Arnold, London. pp. 54.

Feng, S. Y., Feng, J. S., and Yamasu, T. (1977). Roles of *Mytilus coruscus* and *Crassostrea gigas* blood cells in defense and nutrition. *Comp. Pathobiol.*, 3, 31-67.

Feng, S. Y., Feng, J. S., Burke, C. N., and Khairallah, L. H. (1971). Light and electron microscopy of the leucocytes of *Crassostrea virginica* (Mollusca: Pelecypoda). *Z. Zellforsch.*, 120, 222-245.

Foley, D. A. and Cheng, T. C. (1972). Interaction of molluscs and foreign substances: The morphology and behavior of hemolymph cells of the American oyster, *Crassostrea virginica*, in vitro. *J. Invertebr. Pathol.*, 19, 383-394.

Foley, D. A. and Cheng, T. C. (1975). A quantitative study of phagocytosis by hemolymph cells of the pelecypods *Crassostrea virginica* and *Mercenaria mercenaria*. *J. Invertebr. Pathol.*, 25, 189-197.

Foley, D. A. and Cheng, T. C. (1977). Degranulation and other changes of molluscan granulocytes associated with phagocytosis. *J. Invertebr. Pathol.*, 29, 321-325.

160   A. MOHANDAS

Hohl, H. R. (1965). Nature and development of membrane systems
    in food vacuoles of cellular slime molds predatory upon
    bacteria. *J. Bacteriol.*, 90, 755-765.

Malech, H. L., Root, R. K., and Gallin, J. I. (1977). Structural
    analysis of human neutrophil migration. Centriole, micro-
    tubule, and microfilament orientation and function during
    chemotaxis. *J. Cell. Biol.*, 75, 666-693.

McDade, J. E. and Tripp., M. R. (1967). Lysozyme in the hemolymph
    of the oyster, *Crassostrea virginica*. *J. Invertebr. Pathol.*,
    9, 531-535.

Mohandas, A. and Cheng, T. C. (1985). Release pattern of amino-
    peptidase from *Biomphalaria glabrata* hemocytes subjected to
    high-level bacterial challenge. *J. Invertebr. Pathol.*, 45,
    298-303.

Mohandas, A., Cheng, T. C., and Cheng, J. B. (1985). Mechanism of
    lysosomal enzyme release from *Mercenaria mercenaria* granulo-
    cytes: A scanning electron microscope study. *J. Invertebr.
    Pathol.* (in press).

Renwrantz, L. R., Yoshino, T. P., and Cheng, T. C. (1979). Size
    determination of leucocytes from the American oyster
    Crassostrea virginica and description of phagocytosis
    mechanism. *Zool. Jahrb. Abt. Allg. Zool. Physiol. Tiere.*,
    83, 1-12.

Rodrick, G. E. and Ulrich, S. A. (1984). Microscopical studies on
    the hemocytes of bivalves and their phagocytic interaction
    with selected bacteria. *Helg. Meeres.*, 37, 167-176.

Ruddell, C. L. (1971). The fine structure of the granular amebo-
    cytes of the Pacific oyster, *Crassostrea gigas*. *J. Invertebr.
    Pathol.*, 18, 269-275.

Schellens, J.P.M., Daems, W. T., Emeis, J. J., Brederoo, P.,
    DeBruijn, W. C., and Wisse, E. (1977). Electron microscopi-
    cal identification of lysosomes. *In* "Lysosomes, A Laboratory
    Handbook", J. T. Dingle (ed.). North-Holland, The
    Netherlands. pp. 147-208.

Swift, H. and Hruban, Z. (1964). Focal degradation as a biological
    process. *Fed. Proc.*, 23, 1026-1037.

Venugopal, B. and Luckey, T. D. (1978). "Metal Toxicity in
    Mammals. 2", Plenum, New York, pp. 409.

Yoshino, T. P. and Cheng, T. C.   (1976).   Fine structural localization of acid phosphatase in granulocytes of the pelecypod *Mercenaria mercenaria*.   *Trans. Amer. Micros. Soc.*, 95, 215-220.

Yoshino, T. P. and Granath, W. O., Jr.   (1985).   Surface antigens of *Biomphalaria glabrata* (Gastropoda) hemocytes: Functional heterogeneity in cell subpopulations recognized by a monoclonal antibody.   *J. Invertebr. Pathol.*, 45, 174-186.

INDEX